War Inventions

A Tank

THESE WEIRD-LOOKING ENGINES ARE LITERALLY MOVING FORTS, AND ARE
THE EVOLUTION OF A PEACEFUL AGRICULTURAL MACHINE FITTED WITH
"CATERPILLAR" WHEELS, THAT IS, A BROAD BAND ENCIRCLES THE DRIVING WHEELS,
AND SO THE WHOLE CONSTRUCTION MOVES AS IT WERE ON ITS OWN
REVOLVING PLATFORM AND IS THUS PREVENTED FROM SINKING INTO THE
SOFT GROUND. THE PRINCIPLE ITSELF IS NOT NEW, AS IT WAS ADAPTED
TO TRANSPORT CARTS DURING THE CRIMEAN WAR.

War Inventions
Warships, Guns, Tanks, Rifles, Bombs &
Other Instruments and Munitions of
Warfare, How They Were Invented
& How They Are Employed

T. W. Corbin

LEONAUR

War Inventions: Warships, Guns, Tanks, Rifles, Bombs & Other
Instruments and Munitions of Warfare, How They Were
Invented & How They Are Employed
by T. W. Corbin

Originally published under the title
The Romance of War Inventions

Published by Leonaur Ltd

Material original to this edition and this editorial selection
copyright © 2011 Leonaur Ltd

ISBN: 978-0-85706-467-7 (hardcover)
ISBN: 978-0-85706-468-4 (softcover)

http://www.leonaur.com

Publisher's Notes

The views expressed in this book are not necessarily
those of the publisher.

Contents

Chapter 1

How Peaceful Arts Help in War

In the olden times warfare was supported by a single trade, that of the armourer. Nowadays the whole resources of the greatest manufacturing nations scarcely suffice to supply the needs of their armies. So much is this the case that no nation can possibly hope to become powerful in a military or naval sense unless they are either a great manufacturing community or can rely upon the support of some great manufacturing ally or neutral.

It is most astonishing to find how closely some of the most innocent and harmless of the commodities of peace are related to the death-dealing devices of war. Of these no two examples could be more striking than the common salt with which we season our food and the soap with which we wash. Yet the manufacture of soap furnishes the material for the most furious of explosives and the chief agent in its manufacture is the common salt of the table.

Common salt is a combination of the metal sodium and the gas chlorine. There are many places, of which Cheshire is a notable example, where vast quantities of this salt lie buried in the earth.

Fortunately it is very easily dissolved in water so that if wells be sunk in a salt district the water pumped from them will have much salt in solution in it. This is how the underground deposits are tapped. It is not necessary for men to go down as they do after coal, for the water excavates the salt and brings it to the surface. To obtain the solid salt from the salt water, or brine as it is called, it is only necessary to heat the liquid, when the water passes away as steam leaving the salt behind.

Important though this salt is in connection with our food, it is perhaps still more important as the source from which is derived chlorine and caustic soda. How this is done can best be explained by means of a simple experiment which my readers can try in imagination with me or, better still, perform for themselves.

Take a tumbler and fill it with water with a little salt dissolved in it. Next obtain two short pieces of wire and two pieces of pencil lead, which with a pocket lamp battery will complete the apparatus. Connect one piece of wire to each terminal of the battery and twist the other end of it round a piece of pencil lead. Place these so that the ends of the leads dip into the salt water. It is important to keep the wires out of the solution, the leads alone dipping into the liquid, and the two leads should be an inch or so apart.

In a few moments you will observe that tiny bubbles are collecting upon the leads and these joining together into larger bubbles will soon detach themselves and float up to the surface. Those which arise from one of the leads will be formed of the gas chlorine and the others of hydrogen.

It will be interesting just to enumerate the names of the different parts of this apparatus. First let me say that the process by which these gases are thus obtained is called electrolysis: the liquid is the electrolyte: the two pieces of pencil lead are the electrodes. That electrode by which the current enters the electrolyte is called the an-ode, while the other is the cath-ode. In other words, the current traverses them in alphabetical order.

Now it is familiar to everyone that all matter is supposed to consist of tiny particles called Molecules. These are far too tiny for anyone to see even with the finest microscope, so we do not know for certain that they exist: we assume that they do, however, because the idea seems to fit in with a large number of facts which we can observe and it enables us to talk intelligibly about them. We may, accordingly, speak as if we knew for a certainty that molecules really exist.

Now when we dissolve salt in water it seems as if each molecule splits up into two things which we then call ions. Salt is not peculiar in this respect, for many other substances do the same when dissolved in water. All such substances, since they can be ionized, are called *ionogens*.

Now the peculiarity about ions is that they are always strongly electrified or charged with electricity.

At this stage we must make a little excursion into the realm of electricity. You probably know that if a rod of glass be rubbed with a silk handkerchief it becomes able to attract little scraps of paper. That is because the rubbing causes it to become charged with electricity. In like manner a piece of resin if rubbed will become charged and will also attract little pieces of paper. A piece of electrified resin and an electrified glass rod will, moreover, attract each other, but two pieces of resin or two pieces of glass, if electrified, will repel each other. This leads us to believe that there are two kinds of electrification or two kinds of electrical charge. At first these two kinds were spoken of as vitreous or glass electricity and resinous electricity, but after a while the idea arose that there was really one kind of electricity and that everything possessed a certain amount of it, the electrified glass having a little too much of it and the electrified resin a shade too little of it. From this came the idea of calling the charge on the glass a *positive* charge and that on the resin a *negative* charge. Recent investigations seem to show that we have got those two terms the wrong way round, but to avoid confusion we still use them in the old way.

It will be sufficient for our purpose, therefore, if we assume that every molecule of matter has a certain normal amount of electricity associated with it and that under those conditions the presence of the electricity is not in any way noticeable. When a molecule becomes ionized, however, one ion always seems to run off with more than its fair share of the electricity, the result being that one is electrified positively, like rubbed glass, while the other is negatively charged, like rubbed resin.

Thus, when the common salt is dissolved in water, two lots of ions are formed, one lot positively charged and the other lot negatively. Each molecule of salt consists of two atoms, one of sodium and one of chlorine: consequently, one ion is a chlorine atom and the other is a sodium atom, the latter being positive and the former negative.

Now the electrodes are also charged by the action of the battery. That connected to the positive pole of the battery becomes positively charged and the other negatively. The anode, therefore, is positive and the cathode negative.

It has been pointed out that two similarly charged bodies, such as two pieces of glass or two pieces of resin, repel each other, while either of these attracts one of the other sort. Hence we arrive at a rule that similarly charged bodies repel each other, while dissimilarly charged bodies attract each other.

Acting upon this rule, therefore, the anode starts drawing to itself all the negative ions, in this case the atoms of chlorine, while the cathode gathers together the positive ions, the atoms of sodium. Thus the action of the battery maintains a sorting out process by which the sodium is gathered together around one of the electrodes and the chlorine round the other. Those ions, by the way, which travel towards the *anode* are called *anions*, while those which go to the cathode are termed cat-ions.

Thus far, I think, you will have followed me: the chlorine is gathered to one place and the sodium to the other. The former creates bubbles and floats up to the surface and escapes. But where, you will ask, does the hydrogen come from, which we found, in the experiment, was bubbling up round the cathode. Moreover, what becomes of the sodium?

Both those questions can be answered together. The sodium ions, having been drawn away from their old partners the chlorine ions, are unhappy, and long for fresh partners. They therefore proceed to join up with molecules of water. But water contains too much hydrogen for that. Every molecule of water has two atoms of hydrogen linked up with one of oxygen, but sodium does not like two atoms of hydrogen: it insists on having one only. Accordingly the oxygen atom from the water, together with one of the hydrogen atoms, join forces with the sodium atom into a molecule of a new substance, a most valuable substance in many manufactures, called Caustic Soda, while the odd atom of hydrogen, deprived of its partners, has nothing left to do but to cling for a while to the cathode and finally float up and away.

The sum-total of the operation therefore is this: when we pass an electric current through salt water, between graphite electrodes, chlorine goes to the anode and escapes, while caustic soda is formed round the cathode and hydrogen escapes. Let us see now how this is applied commercially.

For the production of Chlorine the apparatus need be little more than our experimental apparatus made large. The anode can be covered in such a way as to catch the gas as it bubbles upwards. In times of peace this gas is chiefly used for making bleaching powder. It is led into chambers where it comes into contact with lime, with which it combines into chloride of lime, a powder which is sometimes used as a disinfectant, but the chief use of which is for bleaching those cotton and woollen fabrics for the manufacture of which this country is famous throughout the world.

The Germans, however, have taught the world another use for chlorine. Those gallant Canadians who were the first victims of the attack by *poison gas* who suddenly found themselves fighting for breath, and a few of whom, more fortunate than the rest, have reached their homes shattered in health with permanent damage to their lungs, those brave fellows suffered from poisoning by chlorine.

We cannot obtain the other product, the caustic soda, by the same simple means. In our little experiment we succeeded in manufacturing some of it in the region around the cathode, and had we drawn off some of the liquid from there we would have been able to detect its presence. But it would have been mixed up with much ordinary salt, and for commercial purposes we need the caustic soda separate from the salt. The principle is, however, just the same, as you will see.

Imagine a large oblong vat divided by vertical partitions into three separate chambers. These partitions do not quite reach the bottom of the vessel, so that there is a means of communication between all three chambers. This is closed, however, by filling the lower part of the vat with mercury up to a level a little higher than the lower ends of the partitions.

Thus we have three separate chambers with communication between them but that communication is sealed up by the mercury.

The two end chambers are filled with salt water, or brine, while the centre one is filled with a solution of caustic soda. In each end compartment is a stick of graphite, both being electrically joined together and so connected up that they form anodes, while in the centre compartment is the cathode.

When the current flows from the anodes it carries the sodium ions with it, just as it did in our little experiment. But its course, this time, is not straight, since in order to travel from anode to cathode it has to pass through the openings in the partitions, in other words through the mercury.

On arrival at or near the cathode the ions of sodium cause the caustic soda to be formed just as in our experiment, but in this case, you will notice, the formation takes place in a chamber from which the salt brine is completely excluded by the mercury.

Brine is continually fed into the outer chambers and the solution of caustic soda is drawn from the centre one, while the chlorine is collected over the anodes.

And now we can go a step further on our progress from common salt to explosive.

In the soap works there are enormous coppers in which are boiled various kinds of fat. The source of the fat may be either animal or vegetable, many kinds of beans, nuts and seeds furnishing fats practically identical with that which can be got from the fat flesh of a sheep, for instance. To this fat is added some caustic soda solution and the whole is kept boiling for some considerable time. This protracted boiling is to enable the soda thoroughly to attack the fat and combine with it, whereby two entirely new substances are formed.

At first the two new substances are not apparent, for they remain together in one liquid. The addition, however, of some brine causes the change to become obvious for something in the liquid turns solid, so that it can be easily taken away from the rest. That solid is nothing else than soap. It remained dissolved in the water which forms part of the liquid until the salt was put in, but as it will not dissolve in salt water, as you will discover if you attempt to wash in sea water, it separates out as soon as the salt is added.

But still a liquid remains: what can that be? It is mainly salt water and glycerine, that sticky stuff which in peace times we put on our hands if they get sore in winter, or take, in a little water, to soothe a sore throat. That it has other and very different uses was brought home to me when, during the war, I tried to buy some at a chemist's, only to learn that it could not be sold except in cases of extreme need under the orders of a doctor.

The mixed liquid is distilled with the result that the water is driven off and the salt deposited, which with other minor purifying processes gives the pure glycerine.

The next step takes us to the explosives factory, where the glycerine is mixed with sulphuric and nitric acids. Now glycerine, as you will have observed, comes from the animal or vegetable sources and therefore is one of those substances known as *organic*, and, like many other of the organic compounds, it consists of carbon, hydrogen and oxygen. Nature has a marvellous way of combining these same three things together in many various ways to form many widely different substances and if, to such a compound, we can add a little nitrogen, we usually get an explosive. Thus, the glycerine, with some nitrogen from the nitric acid, becomes nitro-glycerine, a most ferocious and excitable explosive, the basis of several of those explosives without which warfare as we know it to-day would be impossible.

CHAPTER 2

Gunpowder and its
Modern Equivalents

The origin of gunpowder appears to be lost in antiquity. At all events it has been in use for many centuries and is still made in many countries.

Most boys have tried to make it at some time or other and with varying degrees of success. Such experiments generally lead to a glorious blaze, a delightfully horrid smell and no harm to anyone, the experimenter owing his safety to his invariable lack of complete success, for although other and better explosives have superseded it for many purposes it is capable of doing a lot of harm when it is well made.

It consists of a mixture of charcoal, sulphur and saltpetre ground up very fine and mixed very intimately together. The mixture is wetted and pressed into cakes and dried, after which it is broken up into small pieces. The precise proportions of the various materials seem to vary a great deal in different countries, but generally speaking there is about 75 per cent of saltpetre (or to give it its scientific name, nitrate of potash), 15 per cent of charcoal and 10 per cent of sulphur.

Now gunpowder, like all explosives, is simply some thing or mixture of things which is capable of burning very quickly. When we light the fire we set going the process which we call combustion, or burning, and, as we know from our own experience, that process causes heat to be generated.

What takes place in the fire-grate is that the carbon of the coal enters into combination with oxygen from the air, the two together forming a new compound called *carbonic acid* gas. There is nothing lost or destroyed in this process, the carbon and oxygen simply changing into the new substance, and could we weigh the gas produced we should find that it agreed precisely with the weight of the carbon and oxygen consumed. For the purpose for which we require the fire, namely, to heat the room, the chief feature about this process is not what is formed in the shape of gas, for that simply goes off up the chimney, but the heat which is liberated. We believe that in some mysterious way the heat is locked up in the coal. Latent is the term we use, which means hidden: in other words we believe that the heat is hidden in the coal: we cannot feel it or perceive it in any way, but it comes out when we let the carbon combine with the oxygen.

Why these two things combine at all is one of those mysteries which may never be solved. We have theories on the subject, but all we really know is that under certain conditions if they be in contact with one another they will combine, apparently for the simple reason that it is their nature so to do.

When we apply the match to the fire all we do is to set up the conditions under which the carbon and oxygen are able to follow their natural instincts, so to speak.

A coal fire, as we all know, burns slowly, for the simple reason that it is only at the surface of the lumps that carbon and oxygen are in contact. If we grind up the coal into a fine powder and then blow it into a cloud, so that every tiny particle is surrounded with air, a spark will cause an explosion. That is how these terrible explosions in coal-pits are caused.

This is sometimes seen on a small scale when one shakes the empty fire-shovel after putting coal on the fire to get rid of the fine dust adhering to it and to save making a mess in the fender. That little cloud of fine dust will often burst into flame like a mild explosion.

We see from this that to make an explosion we require fuel, just as we do to make a fire: but we need that it shall be very intimately mixed with oxygen, so that all of it can burn up in practically a single instant. Now in gunpowder we get these conditions fulfilled. We have the carbon in the shape of charcoal, we also have some sulphur which likewise burns readily, and we have saltpetre which contains oxygen.

Thus, you see, we do not need to go to the air for the oxygen, for the gunpowder possesses it already, locked up in the saltpetre. Moreover, we can see now why it is so important for all the materials to be ground up very fine, for it is only by so doing that we can ensure that every particle of charcoal or sulphur shall have particles of saltpetre close by ready to furnish oxygen at a moment's notice.

Another thing to be observed, for it lets us into the great key to the manufacture of nearly all explosives, is the scientific name of saltpetre. It is *nitrate of potassium*, and all substances whose names begin with *nitr* contain nitrogen: while the termination *ate* signifies the presence of oxygen. We need the oxygen to make the explosion but we do not need the nitrogen, yet the latter has to be present for without it the oxygen would be too slow in getting to work.

Nitrogen is one of the strangest substances on earth. Extremely lazy itself, it has the knack of hustling its companions, particularly oxygen, and making them work with tremendous fury. Whenever we get the lazy gas nitrogen to enter into a combination with other things we may confidently look for extraordinary activity of some sort.

So when we put a light to a quantity of gunpowder we set up those conditions under which the carbon and oxygen can combine, and at the same moment our lazy friend the nitrogen turns out his partner oxygen from the nitrate in which they were till then combined and a sudden burning is the result. The solid gunpowder is suddenly changed into a volume of hot gas 2500 times as great. That is to say, one cubic inch of gunpowder changes suddenly into 2500 cubic inches of gas. That sudden expansion to 2500 times its volume is what we term an explosion. If it takes place in an enclosed space so that the gas formed wants to expand but cannot, the result is a pressure of about forty tons per square inch.

If that enclosed space were the interior of a gun, that force of forty tons per square inch would be available for driving out the projectile.

In the early twentieth century gunpowder is still used for sporting purposes and also for some special purposes in warfare, but it has the great disadvantage that it makes a lot of smoke, so that the enemy would be easily able to locate the guns were it to be used in them. As we know so well, by the messages from France, guns and rifles drop their shells and bullets apparently from nowhere and are extremely difficult to locate. That is owing to the use of improved powders one of the great features of which is their smokelessness.

The reason why gunpowder makes a dense smoke, is because the burning which takes place is very incomplete. Therefore, by some such means as a more intimate mixture of the materials a better and more complete burning must be brought about.

One of the best known of the new powders (they are all spoken of as powders, whatever their form, since they have taken the place of the old gunpowder) is nitro-glycerine, the basis of which is glycerine.

The way in which we obtain this useful material has already been explained. It consists of carbon, a lot of hydrogen and some oxygen. These are not merely mixed together but are in combination, just as oxygen and hydrogen are combined in water. Carbon and hydrogen will both combine with oxygen and will give off heat in the process, but in glycerine they are already happily united together and so glycerine itself is no use as an explosive. If, however, we bring nitric acid and sulphuric acid into contact with it a pair of new partnerships is set up, one being water and the other a compound containing carbon and hydrogen, a lot of oxygen and, most important of all, some of that disturbing, restless though lazy nitrogen.

This is nitro-glycerine, a particularly furious explosive, for that curious nitrogen seems to be so uncomfortable in his new surroundings that at the smallest provocation he will break up the whole combination and then there will be a mass of free atoms of carbon, hydrogen and oxygen, all seeking new partners, just right for a glorious explosion.

So furious and untamed is this stuff that it was almost useless until the famous Nobel hit upon the idea of taming it down by mixing it with an earth called Kieselguhr, which reduces its sensitiveness sufficiently to make it a very safe explosive to use. To this mixture Nobel gave the name of dynamite.

It is interesting at this point to compare the action of this typical modern explosive with that of the older gunpowder. The latter is only a mixture: the former is a chemical compound. The smallest particle of material in the gunpowder is a little lump containing millions of molecules and still more of atoms: when the nitrogen has broken up the original nitro-glycerine, just before the explosion actually takes place, we have a mixture of *single atoms*. Thus the mixture is far more intimate in the latter case and the burning is therefore quicker and more thorough.

Another well-known explosive is gun-cotton. Surely this must be a fancy name, for what can harmless, simple cotton have to do in connection with guns. It is a perfectly genuine descriptive name, however. It seems very strange at first, but it is perfectly true that nitrogen, as it turned glycerine into dynamite, can also turn cotton into gun-cotton. Cotton consists mainly of cellulose, a compound of carbon, hydrogen and oxygen, happily combined together and therefore showing, as we well know from experience, no tendency whatever to change into anything else, least of all to *go off bang*. But that state of things is very much changed when we have induced nitrogen to take a hand in the game.

In actual practice, cotton waste, pure and clean, is dipped into a mixture of sulphuric and nitric acids whereby the cellulose becomes changed into nitro-cellulose, just as a similar process changes glycerine into nitro-glycerine. The whole process of manufacture is of course far more than that simple dipping, but that is the fundamental

MACHINE-GUN VERSUS RIFLE.
THIS ILLUSTRATES THE RAPIDITY AND ACCURACY WITH WHICH THE MODERN RIFLE
CAN BE USED. SERGEANT O'LEARY, V.C., TACKLED A GUN CREW OF FIVE AND KILLED
THEM ALL BEFORE THEY HAD TIME TO SLEW THEIR GUN ROUND—A STRIKING
CONTRAST TO THE BROWN BESS OF A HUNDRED YEARS AGO.

fact of it all. The rest is concerned with getting rid of the superfluous acid, tearing the stuff into pulp and pressing it into blocks. It is probably the safest of explosives, since it can be kept wet, in which case the danger of an accidental explosion is practically nil, provided reasonable care be taken. Even when dry, it behaves in a very kindly way. If hit with a hammer, it only burns for a moment just at the point struck. If ignited with a red-hot rod, it burns but does not explode, unless it is enclosed. The burning, that is to say, is not sufficiently rapid to constitute an explosion.

On the other hand, if it be exploded by a detonator, by which is meant a small quantity of a very powerful explosive, such as fulminate of mercury, fired close to it, it then goes off with a violence which leaves little to be desired.

It would be better still could we persuade a little more oxygen to enter into its composition, for as it is there is not quite enough to burn up the other matters completely. That, however, does not cause smoke, since the combustion is complete enough to change everything into invisible gases. With more oxygen more heat might be generated and the power of the explosion be made greater. Still, even as it is, the explosion of gun-cotton has been estimated by a high authority to produce a pressure of 160 tons per square inch, four times as much as gunpowder. Nitro-glycerine has the advantage of a rather larger proportion of oxygen to carbon, resulting in its being rather more energetic.

Yet another class of explosive is made from Coal Tar. This is a by-product in the manufacture of gas for lighting and also in the manufacture of coke for industrial purposes. It comes from the retorts along with the gas in a gaseous form but condenses into a black liquid in the pipes and more particularly in an arrangement of cooled pipes called a condenser specially placed to intercept it.

In the chemist's eyes it is the most interesting of liquids, for it is full of mysteries and possibilities. The most wonderful achievements of chemistry have it for their raw material and there is still scope for much more in the same direction.

If the tar be gently heated in a closed vessel it will evaporate and the vapour can be led to another vessel, there cooled and converted back into a liquid. This looks rather like doing work for nothing, but the various liquids, of which tar is a mixture, evaporate at different temperatures, so that this furnishes a means of separating them. The first liquid thus procured is known as coal tar naphtha, and if it be again distilled it can be subdivided further, the first liquid separated from it being known as Benzine. This, again, is another of those almost numberless things which consist of carbon and hydrogen. Also, like the other similar substances which we have been discussing, it can, if treated with nitric acid, be made to take into partnership a quantity of oxygen and nitrogen.

Thus we get nitro-benzene. We can repeat the process, when it will take more and become *di-nitro-benzene*. Again we can repeat it, thus producing *tri-nitro-benzene*.

The second liquid separated from coal tar naphtha is called Toluene, which again is composed of carbon and hydrogen in slightly different proportions. Like its *confrère* benzene it, too, can be treated with nitric acid, becoming nitro-toluene and then *di-nitro-toluene* and finally tri-nitro-toluene, the deadly explosive of which we read in the papers as T.N.T.

After the naphtha has been removed from the tar another substance is obtained called Phenol, which in a prepared form is familiar to us all as the disinfectant Carbolic Acid. It also can be treated with nitric acid, to produce tri-nitro-phenol, otherwise known as Picric Acid, which after a little further treatment becomes the famous *Lyddite*.

Most of the actual explosives used in warfare are prepared from one or more of the above-mentioned compounds. For example, nitro-glycerine and gun-cotton, having been dissolved in acetone (another compound of carbon, hydrogen and oxygen) and a little Vaseline added, form a soft gelatinous substance which on being squeezed through a fine hole comes out looking like a cord or string, and hence is called Cordite.

Other explosives are finished in the form of sheets, the dissolved gun-cotton or whatever it may be being rolled between hot rollers which give it the convenient form of sheets and at the same time evaporate the solvent.

By combining these various substances various characteristics can be given to the finished explosive. For instance, the one which drives the shell from the gun, known as the propellant, must not be too sudden in its action. It must push steadily. Its purpose is to drive the shell not to burst the gun, wherefore its action must be comparatively slow and continuous so long as the shell is still in the gun. It must *follow through* as the golf player would put it.

The charge in the shell, however, needs to go off with the greatest possible violence so as to blow the shell to pieces and to scatter the fragments so that they do the maximum of damage.

Those explosives, whose function is thus to burst with a sudden shock, are called High Explosives, as distinguished from the propellants which produce a more or less sustained push.

The great fundamental principle which enables large quantities of these powerfully explosive substances to be handled with comparative safety involves the use of two different substances in combination. That which is used in quantity and which actually does the work is made comparatively insensitive, indeed in some cases it is very insensitive, so that it can safely travel by train, by ship and by road and also may be handled by the soldiers and sailors with very little risk. Some of these compounds can be struck or set on fire with impunity. They are none the less violent, however, when, by the agency of a suitable detonator they are caused to explode.

The detonator, of course, has to be very sensitive indeed, but it need only be used in very small quantities, so that by itself it, too, is comparatively safe. Fulminate of mercury is often employed for this purpose—a compound based upon mercury but in which nitrogen of course figures largely.

Thus, there are two things necessary for the successful explosion, one of which is powerful but insensitive, while the other is highly sensitive but relatively harmless since it is never allowed to exist in large quantities, and as far as possible these are kept apart until the last moment.

One other thing may be mentioned in regard to this matter which is of the greatest importance. That is the necessity for the utmost uniformity in these various compounds, so that when the gunners put a charge into a gun they can rely upon it to throw the shell exactly as its predecessor did. Modern artillery seeks to throw shell after shell within a small area which would clearly be quite impossible if one charge were liable to be stronger or weaker than another, for we can easily see that the more powerful the impetus given the farther will the shell go.

To secure this uniformity the greatest care is taken at all stages of the manufacture, and various batches of the same stuff are tested and mixed, and any of them turning out a little too strong are placed with some a little too weak, so that their faults may neutralize each other. By such methods as these a remarkable degree of uniformity is attained, the result of which we see when we read in the papers of the wonderfully accurate gunnery of which our soldiers and sailors are capable.

In conclusion, a word of warning may be appropriate. Reference has been made above to the safety of modern explosives in the absence of the detonators, but do not let that lead anyone to take liberties. Should any reader come into possession of any of these materials, even in the smallest quantities, let him treat it with the utmost respect, for although what has been said about safety is quite correct, it only means comparative safety, there can be no absolute safety where these substances are concerned.

CHAPTER 3

Radium in War

When we remember how all forms of scientific knowledge were called upon to help in the great struggle, it is not surprising to hear that, although in a comparatively humble way, Radium has had to do its share.

Now radium is one of the most, if not actually the most, remarkable substance known. About a generation ago scientific men, or some of them at all events, were getting rather cocksure. Of course they were quite right when they realized how much was known about things and what great strides had been made during the years through which they had lived. They were proud of the achievements of their scientific friends, for I am not imputing personal vanity to anyone, and they had reason to be proud. They made the mistake, however, of thinking that in one direction at least they had learnt all that there was to be known. The present generation of scientific men seem to be almost too prone to go to the other extreme and to dwell rather much on how little we know now and the wonderful things which are going to be discovered in time.

But that is by the way. A generation ago men seem to have pretty well made up their minds that they knew all about atoms. They said that everything was made up of atoms, that the atoms could not be subdivided nor changed into anything else except temporarily by combination with other atoms, and that when these combinations were broken up the atoms remained just as before, quite unchanged. They believed that the atoms were unchangeable and everlasting. Professor Tyndall, in a famous address, referred to this in somewhat flowery language, telling his hearers that the atoms would be still the same when they and he had *melted into the infinite azure of the past*, which a wag translated into the slang expression of the time, *till all is blue*.

Now not very long after Professor Tyndall made this historic speech Professor Henri Becquerel, of Paris, was trying some experiments with phosphorescent materials, that is, materials which glow in the darkness. In the course of these experiments he used some photographic plates upon which, to his surprise, he found marks which he thought ought not to have been there. Thinking at first that he had accidentally *fogged* his plates, as every photographer has done at some time or other, he tried his experiments again with special care but still he got the mysterious marks.

Those marks were caused by some of those *unchangeable and everlasting* atoms deliberately and of their own accord blowing themselves to bits.

For the celebrated Frenchman was not content to let the matter of those mysterious marks rest: he wanted to know what caused them and he did not desist until he was on the track of the secret. It appeared after careful investigation that they were made by the action of something in some of the ore of the metal *uranium* which he had been using. Moreover, this something evidently had the power of penetrating through the walls of the dark-slide to the plate within. Finally, it was tracked down to the uranium itself which was unquestionably proved to be giving off something in the nature of invisible light, or at all events invisible rays, of strange penetrative power. A little later it was observed that certain ores of uranium seemed to give off these rays more freely than would be accounted for by the amount of uranium present, from which fact it was inferred that there must be something else present in the ore capable of giving off the rays much more powerfully than uranium can. Madame Curie ultimately found out two such substances, one of which she called, after her native land, Polonium (for she is a Pole), and the other Radium. It is the latter which is responsible for by far the greater part of the rays formed.

The rays are invisible, but they affect a photographic plate in the same way that light does. They also make air into a conductor of electricity and if allowed to impinge upon a surface coated with a suitable substance they cause it to glow.

This spontaneous giving off of rays is now spoken of by the general term of *radioactivity*, and it has grown into an important branch of science. A number of other substances have been found to exhibit the same peculiar ray-forming powers, notably Thorium, one of the components of the incandescent gas mantle by the prolonged application of a fragment of which to a photographic plate an impression can be obtained due to the rays.

What, then, are these rays? It is found that they are of three kinds, not that they vary from time to time, but that they can be sorted out into three different sorts of rays which are given off simultaneously all the time. The first sort are stopped by a sheet of paper, the second passing easily through a thick metal plate, while the third appear to be identical with X-rays.

For convenience the three sorts are termed Alpha, Beta and Gamma rays, respectively, after the first three letters of the Greek alphabet.

Further, the Alpha rays prove to be a torrent of tiny particles about the size of atoms, indeed if they be collected the gas Helium is obtained, so that evidently they are helium atoms, and since that is one of those substances whose molecules consist of a single atom each they are also molecules of helium. No doubt the reason why they are so easily stopped by a piece of paper is because being complete atoms they are large, huge indeed, compared with the particles which form the Beta rays, for they are apparently those same electrons which are found in the X-ray tube, and which are at least 2000 times smaller than the smallest atom.

When the electrons in the vacuum tube are suddenly brought to a standstill X-rays are given off and in like manner X-rays no doubt would be given off when they start on their journey, providing that they started suddenly enough. Hence it is the starting or sudden explosion-like ejection of the Beta particles which is believed to give rise to the Gamma rays.

The strength or intensity of the rays can be measured very conveniently by their action in making air conductive to electricity, for which purpose a very beautiful but simple instrument called an Electroscope is employed. It consists generally of a glass-sided box or else a bottle with a large stopper, consisting of sulphur or some other particularly good insulator. Through this a wire passes down into the inside of the vessel terminating in a vertical flat strip to the upper end of which is attached a similar strip of gold leaf or aluminium foil. Normally the leaf hangs down close to the strip, but if the wire above the stopper be electrified by touching it with a piece of sealing-wax rubbed lightly against the coat sleeve the charge of electricity passes down into the inside and causes both strip and leaf to become so electrified that they repel each other.

Owing to the non-conductivity of the air in its normal condition the leaf will, if the insulation of the stopper be good, remain projecting almost horizontally for some time until, as it loses its charge by a slow leakage, it gradually settles down close to the strip.

If, however, a piece of radium be brought near while it is sticking out, the leaf will fall almost instantly. X-rays have a similar effect even from several feet or yards away.

The intensity of the radio-activity of different substances can be compared by noting the difference in the rate at which the leaf falls under the influence of each.

What is happening, then, to the atoms of radium, which causes them to show these curious effects and to give off these strange rays? To give any intelligent answer to that question we are bound to assume that which the older generation of scientists thought impossible, namely, that atoms can be broken up. Then we are forced to believe that the atoms of this particular substance radium are of a peculiarly flimsy unstable sort, so that they cannot permanently hold their parts together but are liable to break up, as far as we can see through their own inherent weakness and under the influence of disruptive forces at work within themselves.

We must remember, however, that the tiniest speck of matter which we can see contains a number of atoms of such a size as to be quite beyond the grasp of our minds. To give a rough idea of it in figures is useless as no one can comprehend the real value of a figure or two followed by probably from a dozen to twenty *noughts*. It is best to content ourselves with the general statement that a speck of matter only just visible to the eye contains an exceedingly vast number of atoms. Of course a speck of radium is no exception to this and we must remember, too, that all of them do not break up at once. Indeed, the number breaking up at any time are actually countable by means of a very simple contrivance and a sensitive electrometer. Consequently, in view of the enormous number present and the comparatively small number breaking up at any moment, it is not surprising to hear that, so it is estimated, the process can go on for an almost indefinite number of years, certainly for hundreds. There are, moreover, certain facts which we need not go into here from which the above fact can be clearly inferred, quite apart from what has been said about the vast numbers of the atoms.

It seems as if the uranium atoms break up first, giving off helium atoms and electrons and leaving an intermediate substance called *Ionium* which in its turn breaks up giving off the same things again and leaving radium. That in its turn goes through a complicated series of changes still giving off the same alpha particles or atoms of helium and electrons until, it is suggested, it finally settles down into the simple commonplace metal lead of which we make bullets and water pipes and such-like ordinary things.

We see then that all through its history—its radio-active history at any rate—this stuff is throwing off atoms of helium at a very high velocity (about 50,000 miles a second), and if it be enclosed in anything this enclosing vessel or substance will be subjected to a continual bombardment by the alpha particles. Now just as a piece of iron gets hot if we hammer it, so the enclosing matter is heated by the continual blows which it is receiving night and day, year in and year out, from the alpha particles.

Consequently the immediate surroundings of a speck of radium are always slightly raised in temperature.

Moreover, if a speck of radium be placed against a screen covered with suitable materials each particle which strikes it will make a little splash of light. At least that is what it looks like when seen through a magnifying glass, but to the naked eye there only appears a beautiful steady glow.

Suppose, then, that instead of putting the speck of radiant matter in front of a screen we mix it up intimately with a fluorescent substance such as sulphide of zinc, we then get the same conditions in a slightly different form. Each particle of the substance serves as a tiny screen which glows every time a particle hits it. Thus is produced a luminous paint which glows by night, suitable for painting the dials of instruments which have to be used in the dark.

No doubt some of my readers will have experienced the strangely mingled delight and horror of seeing a Zeppelin in the night sky intent on dropping murder and death on the sleeping civilians of a peaceful town or city. Some too may have witnessed the later acts in that wonderful drama, when, beside the silvery monster illuminated by the beams of the searchlight there must have been, though quite invisible, a little aeroplane manned by one man or at most two. That aeroplane was, no doubt, fitted with instruments at which the pilot glanced now and then and which he was able to see and read because of the tiny speck of radium mixed into the paint. The little alpha particles gave him the light by which to see, but they gave no help to the Germans on the Zeppelin. Hence, in due time he did his work and the gigantic balloon, the pride of the Kaiser and his hordes, fell to the ground, a blazing wreck. How he did it I cannot tell, but of this I am sure, that most probably radium helped him by making luminous and visible the instruments which guided him.

But probably it has rendered and will still render us even greater services in the way of helping to repair the damages to our injured manhood. How many men came back from the war crippled with rheumatism because of the hardships through which they went. That disease is believed to be due to a substance which mingles with the blood and which, although usually liquid and harmless sometimes changes into a solid and settles in the joints. Now it is believed that radium properly administered will act upon that solid and cause it to change back into its liquid form again, thereby curing the disease. Certainly many of the mineral springs at such places as Bath and Buxton give forth a water which shows a certain amount of radio-activity and it may be that which gives those waters their healing properties. If so, we may look forward with confidence to the time when radio-activity will be induced to play a still more successful part in meeting this painful and widespread illness.

Then, of the other ills which will inevitably arise in our men through the hardships which they have endured are sure to be some of the cancerous type, many of which appear to succumb to treatment by radium. If a very small quantity indeed be carried for a few days in a pocket it will imprint itself upon the skin beneath as if it burnt the tissues. It is never advisable, therefore, to carry radium in the pocket without special precautions. One cannot help feeling, however, that in that little fact is a hint of usefulness when the best modes of application have been discovered, for as a means of safely and painlessly burning away some undesirable growth it would seem to be without a rival. It is said, too, that it has the strange power of discriminating between the normal and the abnormal, attacking the latter but leaving the former, so that when applied, say, to some abnormal growth like cancer it may be able to remove it without harmful effect upon the surrounding tissues.

Of this, however, it is too soon to write with confidence. It has not been known long enough for our doctors to find out the best modes of use, but that will come with time: meanwhile there are indications that in all probability it will render good service to mankind.

CHAPTER 4

A Good Servant,
Though a Bad Master

One morning during the war the whole British nation was startled to learn that Mr. Lloyd George, then the Minister of Munitions, had taken over a large number of distilleries. Could it be that he, a teetotaller and temperance advocate, was going to supply all his workers with whiskey? Or was he going to close the places so as to stop the supply of that tempting drink? Neither of these suggestions was his real reason. What he wanted the distilleries for was to make alcohol for the war, not for drinking purposes but for the very many uses which only alcohol can fulfil in most important manufactures.

Probably alcohol is the next important liquid to water. For example, certain parts of shells have to be varnished and the only satisfactory way to make varnish is to dissolve certain gums in alcohol. The spirit makes the solid gum for the time being into a liquid which we can spread with a brush, yet, after being spread, it evaporates and passes off into the air, leaving behind a beautiful coating of gum. That is how all varnishing is done, the alcohol forming the vehicle in which the solid gum is for the moment carried and by which it is applied. It is far and away the most suitable liquid for the purpose, and without it varnishing would be very difficult and unsatisfactory. Hence one need for alcohol, to carry on the war.

Then again some of the most important explosives are solid or semi-solid, and yet they require to be mixed in order to form the various *powders* in use by our gunners. The best way to bring about this mixture is to dissolve the two components in alcohol, thereby forming them both into liquids which can be readily mixed. Afterwards the alcohol evaporates; indeed, one of its great virtues for this and similar purposes is that it quietly takes itself off when it has done its work like a very well-drilled servant.

28

What then is this precious liquid and how is it produced? In order to answer that question it is necessary first to state that there are a whole family of substances called *alcohols*, all of which are composed of carbon, hydrogen and oxygen in certain proportions. There are also a number of kindred substances also, not exactly brothers but first cousins, so to speak, which because of their resemblance to this important family have names terminating in *ol*.

They owe their existence to the wonderful behaviour of the atoms of carbon. In order to obtain some sort of system whereby the various combinations of carbon can be simply explained chemists picture each carbon atom as being armed with four little links or hooks with which it is able to grapple, as it were, and hold on to other atoms. Each hydrogen atom, likewise, has its hook, but only one instead of four.

Now it is easy to picture to ourselves an atom of carbon in the middle with its hooks pointing out north, south, east and west with a hydrogen atom linked on to each. That gives us a picture of the molecule of Methane, the gas which forms the chief constituent of coal gas such as we burn in our homes. Methane is also given off by petroleum and it is the cause of the explosions in coal mines, being known to the miners as *firedamp*. It is the first of a long series of substances which the chemist called paraffins. The first, as you see, consists of one of carbon and four of hydrogen. Add another of carbon and two more of hydrogen and you get the second *Ethane*. Add the same again and you get the third, *Propane*, and so on until you can reach a substance consisting of thirty-five parts of carbon and seventy-two parts of hydrogen. All we need trouble about, however, is the first two, *Methane* and *Ethane*.

We have pictured to ourselves the molecule of methane: let us do the same with ethane. Imagine two carbon atoms side by side linked together or hand in hand. Each will be using one of its hooks to grasp one hook of its brother atom. Hence each will have three hooks to spare on to which we can hook a hydrogen atom. Thus we get two of carbon and six of hydrogen neatly and prettily linked up together. The atoms form an interesting little pattern and to build up the various paraffin molecules with a pencil and paper has all the attractions of a puzzle or game. All you have to do is to add a fresh atom of carbon alongside the others and then attach an atom of hydrogen to each available unused hook. If you care to try this you will get the whole series, each one having one atom of carbon and two of hydrogen more than its predecessor.

If you mix together a quantity of methane and an equal quantity of chlorine, which I have shown you in another chapter how to get from common salt, a change takes place, for in each molecule of methane one hydrogen atom becomes detached and an atom of chlorine takes its place. How or why this change occurs we do not know. It is a fact that the chlorine has this power to oust the hydrogen and there we must leave it, for the present at any rate. The substance so formed is called *methyl* chloride.

In another chapter reference has been made to that substance which is made from common salt and which is so important in so many manufactures called caustic soda. If we bring some of it into contact with the *methyl* chloride the chlorine is punished for its rudeness in displacing the hydrogen; it is paid back in its own coin, for it is in turn displaced not this time by a single atom but by a little partnership called *hydroxyl* one atom of hydrogen and one of oxygen acting together. We can again form a neat little picture of what happens. The oxygen atom has two hooks, one of which it gives to its friend the hydrogen atom and thus they go about hand-in-hand, the oxygen having one unused hook with which to hook on to something else. In this case it hooks on to that particular hook from which it pushes the chlorine.

We have thus seen two changes take place. First, the hydrogen is displaced by the chlorine: then the chlorine is turned out and its place taken by the *hydroxyl*. And during both these changes the central carbon atom and its three hydrogen partners have remained unaffected. Those four atoms are called the *methyl* group, and a *methyl* group combined with a *hydroxyl* group forms *methyl alcohol*.

Similar changes can be brought about with Ethane as with Methane, and in them the two carbon atoms and the five hydrogen remain unchanged, whence they too are regarded as a group, the *Ethyl* group, and an *ethyl* group hooked on to a *hydroxyl* group gives us a molecule of *ethyl alcohol*.

These groups of which we have been speaking never exist separately except at the moment of change, but in the wonderful changes which the chemist is able to bring about the atoms forming these groups seem to have a fondness for keeping together and moving together from one substance into another. In a word, they behave as if they were each a single atom and they are called by the name of radicles; the word simply means a little root.

The *methyl* radicle and the *ethyl* radicle, since they form the basis of two of the paraffin series, are called paraffin radicles, so that we can describe this useful alcohol as a paraffin radicle with a *hydroxyl* radicle hooked on to it. If we use the *methyl* radicle we get *methyl* alcohol: if we use the *ethyl* radicle we get *ethyl* alcohol.

Now *ethyl* alcohol is the spirit which is contained in all strong drink. Whiskey has as much as 40 per cent and brandy and rum about the same, while ale has only about 6 per cent. All of them may be regarded as impure forms of *ethyl* alcohol, the various impurities giving to each its particular taste.

Ethyl alcohol, too, is what is sold at chemists' shops as *spirits of wine*, where also we can purchase that which is familiar as *methylated spirits*, whereby there hangs a tale.

All Governments regard alcohol for drinking as a fit subject for taxation. When anyone buys a drink with alcohol in it a part of what he pays goes to the Government in the form of duty. On the other hand, when alcohol is used for trade purposes, for making varnish or something like that, there is no reason whatever why it should be charged with duty. But if the varnish manufacturer is to have alcohol duty-free what is to prevent him from using some of it for drinking?

To get over the difficulty, that which is supplied to him or to anyone else for trade purposes is deliberately adulterated so as to make it so extremely nasty that no one is likely to want to put it in his mouth.

It so happens that *methyl* alcohol, while as good as the other for many purposes, is horrible to the taste and so it forms a very convenient adulterant for this purpose. Therefore, when methylated spirit is sold to you for drying your photographs, the chemist gives you *ethyl* alcohol with enough *methyl* alcohol in it to make sure that neither you nor anyone else will ever want to drink it.

That, then, is alcohol: a near relative of paraffin oil and also of coal gas, yet it is from neither of these that we get it. The changes described above enable you to realize what it is, but they do not tell how it is made in large quantities.

Ethyl alcohol is obtained from sugar by the employment of germs or microbes. Any sort of sugar will do: it need not be sugar such as we eat. In practice the sugar is usually obtained from starch, that very common substance which forms the material of potatoes, grain of all kinds, beans and so on. There is a kindly little germ which will quite readily turn starch into sugar for us if we give it the chance.

The maltster starts the process. He gets some grain, and spreading it out in a damp condition upon his floor sets it a-growing. As soon as it has just started to grow, however, he transfers it to his kiln, where by heating it he kills the young plants. As is well known, every seed contains the food to nourish the little growing plant until it is strong enough to draw its supplies from the soil and the food thus provided for the young wheat plant is starch, which, when it is ready for it, it turns into sugar. The little shoot lives on sugar and the maltster and distiller conspire to steal that sugar intended for the baby plants and turn it into alcohol.

So the little plant liberates by some wonderful means a material called diastase, which has the power of changing starch into sugar. It does it, of course, for the purpose of providing its own necessary food, but the maltster does not want the process to go too far: he only wants to produce the diastase, and that is why he kills the plants, after which he has finished with the matter and hands the *malted* grain or *malt* over to the distiller for the next process.

The distiller mixes the malt with warm water, whereupon the diastase commences the conversion of the starch of the grain. At this stage fresh grain may be added and potatoes, indeed almost anything composed largely of starch for the diastase to work upon. The process goes on until, in time, the liquid consists very largely of sugar dissolved in water, which is strained away from what is left of the grain, etc.

Malt sugar is very similar to, but not quite the same as, cane sugar. It consists of twelve parts of carbon, twenty-two of hydrogen and eleven of oxygen. It is an interesting little puzzle to sketch those atoms out on paper, each with its proper number of hooks, and see how they can be combined together. Malt sugar, milk sugar and cane sugar all consist of the same three elements in the same proportions and the difference between them is no doubt due to the different ways in which the atoms can be hooked up together.

Yeast is next added to the liquid, upon which the process of fermentation is set up, the tiny living cells of the yeast plant producing a substance which is able to change the sugar into alcohol.

The alcohol thus formed is, of course, combined with water, but it can be separated from it by gentle heating since it passes off into vapour at a lower temperature than does water. Thus the vapour first arising from the mixture is caught and cooled whereby the liquid alcohol is obtained. This operation, called fractional distillation, has to be repeated if alcohol quite free from water is required, in addition to which the attraction which quicklime has for water is called into play to coax the last remnant of water from the other.

And now, how about the *methyl* alcohol? That is obtained in quite a different way, by heating wood and collecting the vapours given off by it. Hence it is often called *wood spirit*.

As a matter of fact, at least two very valuable substances are obtained by this operation, *methyl* alcohol and acetone.

The vapours given off by the wood are cooled, whereupon tar is formed while upon it there floats a dark liquid which contains the wood spirit, acetic acid and acetone.

To capture the acetic acid lime is added to the mixture, and since there is a natural affinity between them, the acetic acid and lime combine into a solid which remains behind when the whole mass is suitably heated. What comes over in the form of vapour is a mixture of water, acetone and wood spirit. The former is enticed away by the use of quicklime, while the other two are separated by the process of fractional distillation already referred to.

Now let me ask you to form another little picture, either in your mind or with paper and pencil. Imagine two *methyl* radicles, each, let me remind you, a carbon atom with three hydrogen atoms hooked on and one spare hook. Also imagine one atom of oxygen with its two hooks outstretched like two arms, and just link one radicle on to each. Then you have the picture of *methyl* ether. All the ethers are formed by taking two of the paraffin radicles and linking them together by means of the two hooks of an oxygen atom. The ether which is so largely used in hospitals for wounded soldiers is *ethyl* ether, consisting of two *ethyl* radicles joined by oxygen. How it is made we will come to in a moment, but as you see already it is a close relative of alcohol.

Now from *methyl* ether take away the central oxygen and in its place put carbon. This atom will have two hooks to spare which it can employ to hold on to the two hooks of the oxygen. The result is a molecule of acetone.

This is used as a solvent in a similar manner to alcohol for many purposes, and there was a great demand for it no doubt during the war.

One interesting use of acetone is in connection with the gas acetylene. Of great use both for lighting and also in conjunction with oxygen for welding and cutting metals, this gas suffers from the disadvantage that it cannot be compressed into cylinders and carried about as oxygen can. It can, however, be dissolved in acetone. The cylinders in which it is carried are therefore filled with coke saturated with acetone and then when the acetylene is pressed in it dissolves, coming out of solution again as soon as the pressure is released. In this dissolved condition it is quite safe to carry about.

For a moment let us turn back to the commencement of the chapter to the subject of methane. When mixed with chlorine, it will be remembered, one hydrogen atom gave place to a chlorine atom. If the process be repeated another hydrogen atom will be displaced in the same way, while a further repetition will result in the removal of a third, when there will be a carbon atom in the centre with three chlorine and one hydrogen hooked on to it. With that picture in your mind's eye you will be contemplating the molecule of that wonderful and beneficent substance, chloroform. When we think of the numberless operations which have been carried out by the surgeons in the course of this last war we realize a little how great is the total sum of pain and suffering which has been saved through the agency of this substance, this simple neat little arrangement of five tiny atoms.

Now that again is obtained in manufacture from alcohol. Alcohol, bleaching powder and water are mixed and then distilled, by which of course is meant that the mixture is evaporated by heat and the vapour collected and cooled back into liquid again. The liquid so obtained is chloroform.

Hardly less important than this, in our military hospitals, is ether, to which reference has already been made. It, too, is manufactured from alcohol. The alcohol, together with sulphuric acid, is placed in a still and heated, the vapour given off being led to another vessel and there condensed. The liquid thus obtained is ether and so long as the supply of fresh alcohol is kept up the production of ether goes on continuously.

The sulphuric acid does not disappear and so does not need to be replaced, from which it would appear as if it might just as well not be there, but that is not the case. It plays the part of what is called a *catalyst*, one of the curiosities of chemistry. There are many instances in which two things will combine only in the presence of a third which appears to be itself unaffected. This third substance is a catalyst. It reminds one of the clergyman at a wedding who unites others but remains unchanged himself.

In conclusion, one may mention that many of the medicines with which our injured men were coaxed back to health and strength owe their existence to alcohol, for many drugs are obtained from vegetable substances by dissolving out a part of the herb with alcohol.

Thus, as a drink, it is unquestionably very harmful. Indeed, in that way it probably kills more people per year than its use in the manufacture of explosives caused in the worst year of the war. Yet it also furnishes chloroform, ether and medicinal drugs and performs a whole host of useful services to mankind. Finally, if oil and coal should ever run short it is quite prepared to run our engines for us. Truly it is a wonderful substance.

CHAPTER 5

Mines, Submarine and Subterranean

The word mine in its military sense originally meant just the same as it does in the ordinary way, but like many other words it has got twisted into new uses the connection of which with the original meaning is very obscure. One of the most striking of these verbal puzzles is the submarine *mine*. There seems at first sight not the remotest connection between the floating barrel of explosives concealed beneath the water and what we ordinarily call a mine. The explanation of this is that the term has acquired this meaning after passing through a series of stages.

When soldiers *mine* for the purpose of blowing up their enemies they dig a hole in the ground, and conceal therein a quantity of explosives so arranged that they blow up when the enemy pass over or near. The operation of digging the hole in the earth is clearly akin to the work of the miner and so such is quite appropriately called a *mine*.

The hole may be dug from the surface downwards, the marks of excavation being afterwards covered up and obliterated as much as possible. In other cases the hole may be a tunnel starting from a trench and driving towards the enemy's position. The idea, of course, is to burrow until the end of the tunnel is just under some important part of the enemy's works or fortifications. When the end of the tunnel has reached the right spot explosives can be placed there, the tunnel partly stopped to prevent the explosion from driving back upon those who make it and the whole fired at the desired moment.

This tunnelling is also called *sapping* and the tunnel itself a sap. Military engineers are often spoken of as *sappers and miners* as if the two things were clearly different, but as a matter of fact both are often used to describe the same thing. Roughly, we may say that a mine which stays still in the hope that the enemy will walk upon it is a mine proper, while a mine which itself progresses towards the enemy until it ultimately goes off beneath him, is a *sap* and the making of such a thing is *sapping*. Or we might say that sapping is undermining, in which sense we use it in general conversation when we speak of something sapping a man's strength. Soldiers speak of their engineering comrades as *sappers* just as they term artillerymen *gunners*, but the only reason why they call them by that name instead of miners is because the latter is a well-known term applied to those who work in coal mines.

A subterranean mine, then, is nothing more or less than a hole in the ground, made in any way that may be convenient, filled with explosives and fired at a suitable time to do damage to the enemy.

In other words, it is simply some explosive *concealed in the ground* with means for firing it, and when the sailor *conceals explosives in the sea* so that they may blow up the enemy's ships, he borrows his military comrades' term and calls it a *mine* too.

Counter-mining is the enemy's reply to mining. Suppose I was foolish enough to wish to blow up my neighbour who lives in the house opposite to mine. I might start from my cellar and dig a tunnel under the road until I knew that I had arrived under his dwelling. But suppose that he got to know of my little scheme: he could then try counter-mining. In this case it would mean starting a tunnel of his own from his cellar towards my tunnel: then, as soon as the two tunnels had come sufficiently near to each other, he could let off his explosives thereby wrecking my tunnel and putting an end to my operations while yet I was only half-way across the road. Thus he would stop me before I had had time to harm him, and since he need only tunnel just far enough to render the necessary explosion harmless to his house, while I to succeed would have to tunnel right across the road, the man who is counter-mining always has a slight natural advantage over the man who is doing the mining. If only he gets to know what is going on in time he can always retaliate.

All forms of land mine are improvised on the spot according to circumstances. Not so, however, with submarine mines on which much ingenuity has been expended, the mines being made in workshops ashore ready for laying and then laid by ships and sometimes by divers.

Of these there are two main kinds, those which are put in place in times of peace for the protection of particular harbours and channels, and those which are simply dropped overboard from a mine-laying ship during the actual war.

They all consist essentially of a case of iron or steel plates riveted together just as a steam boiler is made, in fact the cases are made in a boiler shop. The charge is gun-cotton fired by a detonator, the latter being excited by a stroke from a hammer, as in a rifle, or else by electricity. In the latter case, a tiny filament of platinum wire is in contact with the detonator, and the wire being heated by the current, just as the filament of a lamp is, the detonator is fired by the heat.

Of the permanent mines whereby the entrances to important channels are protected arrangements are often made for firing by observation, that is to say, by the action of an observer ashore. Being laid by divers and securely anchored to heavy weights laying on the bottom, wires are carried from the mines to the observation station. The observer watches and fires the mines at the right moment by simply pressing a key thereby making the electrical circuit.

More often, however, mines are fired by contact. Observation mines have the advantage that while they may be exploded under an enemy they will allow a friendly ship to pass in perfect safety. Contact mines, on the other hand, will afford protection against attacks by night when enemy craft may attempt to creep in under cover of darkness.

Contact mines are often fired electrically, sometimes by batteries of their own inside their own cases, or else by current from the shore through wires, the circuit being completed by an automatic device of some sort actuated unwittingly by the unfortunate victim.

One of these contact devices will illustrate the general character of them all. Imagine a little vessel with mercury in it: it is, generally speaking, of some insulating material, but right at the bottom is a metal stud with which the mercury makes contact. The rim may likewise be of metal or a metal rod may project downwards into it: it matters not which, for we can see at once that it is quite easy so to arrange things that whereas, while upright, the mercury shall be well clear of the upper contact, it shall when the vessel is tilted flow on to it, thereby bridging from lower contact to upper contact and completing the circuit.

Of course, a mine must only go off when actually struck by a ship and not when it is gently swung to and fro by the action of tide or current in the water. That is easily arranged, for the vessel and contacts can be so shaped that contact is not made until an angle of tilt is reached which no tide or ordinary commotion of the water could bring about.

It is clearly possible, too, to combine the contact and observation arrangements in such a way that contact mines can be made safe for friendly ships during the daytime. It is only necessary to adopt the shore battery arrangement already mentioned and disconnect the

AN ITALIAN MINE-LAYER

THIS PHOTOGRAPH WAS TAKEN LOOKING DOWN UPON THE DECK OF THE SHIP. THE
MINES RUN UPON RAILS, AND ARE PUSHED BY THE MEN TOWARDS THE STERN,
WHENCE THEY ARE DROPPED ONE AT A TIME INTO THE WATER. THE SPLASH
INDICATES THAT ONE HAS JUST FALLEN.

batteries during the day or when no enemy is in sight, restoring the connection during the darkness or in the event of hostile ships trying to rush the passage.

Another interesting scheme for keeping mines safe until required is to anchor them in what is termed a *dormant* condition. This means that a loop is taken in the wire rope by which they are anchored, the loop being fastened by means of a link. This link, however, contains a small quantity of explosive which can be fired from the shore. This has the effect of breaking the link, releasing the loop and allowing the mine to float upwards to the full length of the rope. Thus the mine is down deep, well below the bottom of the biggest ship until released for action.

It is doubtful whether much use is made nowadays of permanent mines of the types just described, for they have, no doubt, been largely displaced by the temporary mine which can be laid in a moment by simply being dropping overboard from a ship, but it is quite possible that some of the defences of, say, the Dardanelles, were of the permanent nature.

So let us pass on to the temporary mines. These were used by the Germans from the first few hours of the war. One of the first naval incidents was when our ships discovered a small German excursion steamer which had been converted into a mine-layer strewing these deadly things surreptitiously in the North Sea in the hope that some of our vessels would run upon them. Needless to say, that ship went on no more excursions.

Laid thus, it is evident that there can be no wires running ashore, so that all mines of this class must be contact mines. What makes them of extreme interest is the way they are laid. Just think for a moment what is involved. From the very nature of things their laying must often be done in secret. It is not the British practice to place them in the open seas, except avowedly, after due notice, in certain specified areas, where they are laid quite openly under the protection of adequate forces to ensure against interruption. There is little doubt, however, that they have laid many a mine field secretly in purely German waters, while everyone knows that the Germans have not hesitated to sow the shipping routes broadcast with these things, such work of course being done secretly and largely at night.

The mine can therefore only be laid by dropping it into the water and leaving it. Yet it must not float on the surface or it will be easily seen and picked up; it must float below, so that the unsuspecting ship may run upon it. And it is quite impossible to make a thing float in water anywhere except upon the surface. If it does not float upon the surface it sinks to the bottom: there is no *half-way house* between. Many people are surprised to hear this, judging, no doubt, by the fact that a balloon floats *in*, and not on, the air and expecting an object floating in water to be able to do the same thing. The difference is due to the fact that air is easily compressible, so that the air close to the earth is denser, more compressed, and therefore heavier, than the air higher up owing to its having the whole weight of the upper air pressing downwards upon it. The density of the air diminishes, for this reason, as one ascends, and a balloon which displaces more than its own weight of air at the surface of the earth rises until it has reached just that height when the air displaced exactly equals in weight the balloon itself: then it goes no higher.

Precisely the same conditions exist in the sea except that water being incompressible is no denser at the bottom of the sea than on the surface. Therefore, if a thing sinks at all it sinks right to the bottom.

There is one very ingenious device for overcoming this difficulty by means of a motor and propeller. The mine has enclosed in its case a motor driven by a store of compressed air which operates a propeller. In this it is somewhat like a torpedo, but in this case the propeller is set vertically so that its action lifts the mine up in the water. Now the mine is so weighted that it just and only just sinks when dropped in, but on reaching a certain depth the motor starts and by means of the propeller raises it nearly to the surface again. On nearing the surface the motor stops and the mine sinks once more, only to be raised again in due course, so that the thing keeps on rising and falling; it never rises above a certain depth nor falls below a certain depth, but oscillates continually between its two limits.

The question then arises, what starts and stops the motor at precisely the right moments to produce this result? It is done by means of a hydrostatic valve. As just pointed out, the water at the bottom of the sea is supporting the weight of all that water which is above it. The water is not compressed by this, but the pressure is there all the same. Obviously the degree of pressure at any point depends upon the weight of the layer of water above, and since the weight of that layer will obviously increase and diminish with its thickness it follows that, starting from the surface, where the pressure is nil, we get a perfectly steady and regular increase as we descend, until we reach the maximum at the bottom. Now within the mine is a small watertight diaphragm, the outer surface of which is in contact with the water and upon which, therefore, the water presses. As the mine descends, therefore, this diaphragm is bent inwards more and more by the pressure of water and that is made to start the motor. Adjustments can easily be made so that a certain degree of bending shall result in starting the motor, which is the same as saying that the motor shall start automatically at a certain depth. Likewise as the mine rises under the influence of the propeller the pressure decreases, the diaphragm straightens out and at a certain predetermined depth the motor is stopped.

When, finally, the store of motive power is exhausted the mine sinks to the bottom and is lost, a very valuable feature from a humanitarian point of view, since it means that the active life of the mine is short and it cannot go straying about the oceans for weeks or even months, finally blowing up some quite innocent passenger ship.

More often, however, this difficulty of depth is overcome by anchoring the mine at the depth most suitable for striking the bottom of a passing ship. But here again there seem to be insuperable difficulties, for the depth of the sea varies and so the length of the anchor rope must be varied with almost every mine that is laid. It has been found possible, however, to make the mines automatically adjust the length of their own anchor ropes so that the desired result is attained without difficulty no matter how deep the sea may be. Let me describe how it is done in the Elia mines used by Great Britain. The inventor, Captain Elia, was an officer in the Italian Navy.

The mine consists of three parts: (1) the mine proper, a case containing the explosive, gun-cotton and the firing mechanism; (2) the anchor; and (3) the weight, all of which are connected together by suitable wire ropes.

The mine is lighter than water and so floats: the anchor, which bears no resemblance to the ordinary anchor but which is an iron case containing mechanism, only able to act as an anchor by virtue of its weight, is heavier than water and so sinks, while the weight of solid cast iron sinks more readily still.

The anchor is often fitted with wheels so that it forms a truck upon which the mine and the weight are placed, the whole running upon rails laid on the deck of the mine-layer. As this ship steams ahead the men push the mines along the rails, dropping them over the stern at regular intervals.

When the thing reaches the water, the weight sinks the most rapidly, thereby tugging at the chain whereby it is connected to the anchor. The latter, being less compact, sinks more slowly so that the pull upon the rope is maintained until at last the weight rests upon the bottom. *Then and only then is the pull relaxed.* Now inside the anchor is a winch, upon which is wound a length of flexible wire rope, the other end of which is attached to the mine. The latter, it will be remembered, is light enough to float and so, since it lies upon the surface while the anchor sinks, the rope is drawn off the winch. But there is a spring catch which is able to hold the winch and to prevent it from paying out rope, and that catch is only held off by the pull of the weight. Consequently, as soon as the weight touches the bottom and its pull upon the anchor ceases, the winch is gripped by the catch, no more rope is paid out, and from that moment, as the anchor descends, it drags the mine down with it.

The result, then, is that the mine becomes anchored at a depth below the surface roughly equal to the length of the rope connecting weight to anchor.

Mines of this kind can, of course, be fired electrically by the tilting of a cup of mercury or similar device as already described. Another arrangement is to fit projecting horns upon the surface of the mine made of soft metal so that they will be bent or crushed by a strong blow such as a passing ship would give. This breaks a glass vessel inside, liberating chemicals which cause detonation.

The method adopted in the Elia mines is to have a projecting arm pivoted upon the top of the mine. The mine is spherical (they are nearly all either spherical or cylindrical), with the rope attached to the South Pole, so to speak, and the arm pivoted to the North Pole. As the mine floats in the water the arm projects out horizontally. The effect of this arrangement is that when a ship strikes the mine the latter rolls along its side, but the arm being too long, simply trails along. Thus the spherical case of the mine turns while the arm remains still and that is made to unscrew and eventually release a hammer which, striking the detonator, fires the mine.

In other words, this type of mine is exploded not by the ship giving it a blow, but by its rubbing itself along in contact with the mine. The great advantage of this is that it is only a ship that can do this. No chance commotion in the water can do it: no chance blow from floating wreckage can do it: only the rubbing action of a ship can accomplish it. Such a mine, too, is less likely to be affected by countermining, of which more presently.

Apparently the laying of these mines must be very dangerous work, for since a blow will explode most of them, what is to prevent their receiving that blow while on the deck of the mine-layer, or at all events as they are dropped into the water.

In all cases, precautions are taken against such an event. Sometimes a hydrostatic valve is employed, the arrangement being that the firing mechanism is locked until released by the valve, until, that is, the mine is immersed to a predetermined depth in the water.

Another device for the same purpose is a lump of sugar. The mine is so made that it cannot be fired until this lump has been melted by the action of the water: *sal ammoniac* is another substance employed for the same purpose. The technical term for this is a *soluble seal*. The firing arrangement, whatever it may be, is sealed up so that it cannot come into operation until the seal has been dissolved away by the water, or until the mine has been in the water long enough for the mine-layer to get out of harm's way.

Another interesting feature of the Elia mine is connected with the source of the power which drives the hammer which causes the explosion. The anchor, it will be remembered, pulls the mine down under water, the latter being of itself buoyant. There is a continual pull, therefore, upon the rope by which the mine is held under. It is that pull which works the hammer.

And now observe the beautiful result of that simple arrangement. Suppose the mine breaks its rope and gets loose, so that it can drift about and carry danger far and wide. It can break loose and it can drift about, but at the very moment of getting loose the danger vanishes, for the rope ceases to pull and the firing mechanism loses its motive power.

In other mines the same result has been sought by means of clockwork, which throws the firing arrangements out of action after the lapse of a given time. This scheme of Captain Elia's, however, whereby the very act of breaking adrift produces its own safeguard, is one of the most delightful instances of a happy invention.

In conclusion, just a word about the measures taken against mines. Counter-mining is one. It consists in letting off other mines in the midst of a mine-field with the purpose of giving them such a shaking up that some of them will be exploded by the shock.

The simplest and indeed the only effective way, however, seems to be the simple primitive method of dragging a rope along between two light draught vessels and thus tearing the mines up by their roots, so to speak. The very act of thus dragging it along by its anchor rope often causes a mine to explode, well astern of the mine-sweeping vessels, but sometimes they are pulled up and fired or sunk by a shot from a gun which the sweeper carries for the purpose.

The sweeping up of the mine-fields is a duty often allotted to the steam fishing boats or trawlers, whose crews seem particularly well fitted for the work. It is a hazardous duty, and many lives have been lost through it. Let us hope that in time to come all submarine mines and the dangers connected with them will be a thing of the past, for they are mean, cowardly and contemptible weapons.

CHAPTER 6

Military Bridges

Bridging has always been an important part of actual warfare. In my school days I studied Caesar from a textbook which is not much in use nowadays and which had very copious notes, prominent among which was a description, with drawings, of a bridge made by the Roman Legions in Gaul. And a fine bridge it was, too. How its details came to be known was partly through the description given by Caesar himself and partly by a study of certain old timbers found in the bed of the Rhone, which timbers were believed to be relics of the very bridge which the great Julius himself had had built.

This bridge of nearly two thousand years ago appeared to be built of *baulks* of timber fastened together in very much the same manner as that adopted by the engineering units of the great armies of to-day.

Every observant person has noticed how tall poles and short sticks tied together with ropes can be fashioned into the firm, strong scaffolding from which workmen can in safety raise great tall buildings. That mode of construction can always be used to form a bridge.

Equally well known, no doubt, are the gantries built over the footway while a large building is in course of construction. Generally of huge square *baulks* of timber, they are intended to carry very heavy loads of materials and to save the public passing beneath from any possibility of damage through heavy objects falling from above. Those gantries furnish us with an example of another sort of construction in wood which can be and is often used in bridging.

When the Germans retired in Northern France they blew up all bridges behind them, and before the Allies could use those bridges they had to repair them. If only for foot-traffic, a contrivance of poles, lashed together after the manner of the builder's scaffold, is ample in most of such cases and by its means a strong and safe bridge can be made upon what is left of the old bridge in the course of a few hours.

For light vehicles a similar structure but made stronger by more lashings and of poles closer together will suffice, but for heavy traffic, with guns and possibly railway trains, recourse has to be had to the heavy timberwork exemplified by the builder's gantry. This takes longer to make, since the timbers are big, heavy and not easy to move about: they are, moreover, not simply laid beside or across each other and tied, but are cut the right lengths, and one is notched where the end of another fits into or against it. The *baulks* are connected by bolts and nuts for which holes have to be drilled or by rods of iron with a sharply pointed prong on each end stretching across from one baulk to another, one prong being driven into each.

With the long-thought-out military operations of modern warfare it is just possible that steelwork for repairing certain particular bridges might be prepared in advance and simply launched across when the time arrives, but that is manifestly impossible except in certain cases and under particularly favourable conditions, such as railway facilities for bringing up the new bridge close to the site where it is to go.

Nearly every military bridge therefore has to be more or less improvised on the spot. In a highly developed country scaffold poles or *baulks* may be found or brought up by road or rail, in less civilised lands their equivalents may be cut and prepared from neighbouring forests, but all armies have, as a recognised part of their organisation, certain engineering *field companies*, and bridging trains, which carry with them large quantities of material carefully schemed out long in advance, so shaped and so prepared that it can be fashioned into almost anything, much as the strips of a boy's Meccano can be adapted to form a great variety of objects.

First, there are pontoons, large though light boats or punts, about 20 feet long, constructed of thin wood with canvas cemented all over to give additional strength and water-tightness. Each pontoon rides upon its own carriage upon which there are also stowed away quantities of timbers of various sorts, anchors for holding the pontoons in place, oars for rowing them, ropes of different kinds, and so on.

Each pontoon, moreover, is divided about the middle into two pieces called respectively the bow piece and the stern piece. The two are normally coupled together by cunningly devised fastenings but they can be quickly separated, in which state they form two shorter boats.

Other carriages carry more timber and material intended for the purpose of forming *trestle bridges* but which is also usable in connection with the pontoons.

Of this material the chief sorts are *legs*, long straight pieces which form the uprights; transomes, heavier beams which can be fitted across horizontally between two legs so that the three form a huge letter H or a very robust Rugby goal; *baulks* which are light timbers tapered off towards each end for the sake of lightness and of such size that they fit snugly into notches which are cut in the upper surface of the transomes; and planks called *chesses* for forming the floors of a bridge.

Probably the most dramatic incident of the war was when the British, having been apparently beaten by the Turks in Mesopotamia, driven far back and their General and many troops captured, suddenly turned the tables upon their enemies, driving them from Kut and sending them fleeing helter-skelter to Bagdad and then beyond. Now the capture of Kut and then of Bagdad were both made possible by the rapid bridging of the Tigris, and without doubt this is the sort of material which was used. Let us see how it is done.

An army arrives at a river across which it is decided to throw a pontoon bridge. The pontoons are unloaded off their wagons and launched into the water. One is rowed out and anchored a little way from the shore, while upon the bank parallel with the river is laid a *transome*. On the centre of the pontoon is a centre beam with notches in it like those in the transomes and from the one to the other *baulks* are passed. Meanwhile a second pontoon has been rowed into place and more *baulks* are passed from the first pontoon to the second, while chesses are laid upon the *baulks* to form a platform or floor.

Thus, pontoon by pontoon, the bridge grows until it has reached the further bank.

If pontoons are scarce and the loads to be carried by the bridge are light they are divided in two, and instead of a row of pontoons joined by *baulks* there is a row of *pieces* joined by *baulks*. Pieces arranged thus form a light bridge, pontoons a medium bridge, while pontoons placed closer together form a heavy bridge. Which shall be built depends upon the number of pontoons available in relation to the width of the river and the nature of the traffic which will have to pass over.

An alternative arrangement is to make the pontoons up first into groups or rafts and then bridge from raft to raft instead of bridging between pontoons.

There is still another way of making the bridge, and that is to put it together alongside the bank, afterwards swinging it across the river like the opening or shutting of a door. Anyone can see that there must be many advantages in this latter method when it is practicable, since more men can work at once and with greater safety, for all will be near the bank.

It is evident that such a structure depends for its security entirely upon the anchors. Those which are carried for the purpose are like those of a ship but there may not be enough or they may not suit every kind of river-bed. They are often improvised therefore. Two wagon wheels lashed together, with heavy stones clipped between them, are said to be a very effective anchor. Under certain conditions a net filled with stones is surprisingly effective. Two pickaxes tied together form a good imitation of the conventional anchor, as also does a harrow sunk and held down by stones thrown upon it.

Trestle bridges are made in quite a different way. The trestles are formed of two legs or uprights with a transome between, a shape which resembles, as has been already remarked, a very robust Rugby goal. The transome is connected to the legs by a special form of band which permits it to be fixed at any height without having to drill any special holes for the connections. The legs are so shaped at their ends that they can be shod with steel shoes provided for the purpose, enabling them to get a good foothold even on shifty soil. The trestles are put together ashore, and each is taken out in a boat or on a pontoon to the place where it is to stand. Then it is launched feet foremost into the water, the boat being on the side away from the shore, so that a rope from the trestle to the shore will enable men on land to pull the trestle into an upright position.

Thus trestle after trestle is added until the bridge has grown right across the water to the further bank. The trestles cannot fall over sideways because of their own width, they cannot fall forwards or backwards because of the *baulks* which pass between them and carry the floor, but as a precaution diagonal ties of rope are always added here and there along the bridge, that is to say, two trestles are tied together with two ropes, each rope passing from the bottom of one trestle to the top of the other, a form of tying which is very effective and very easy and simple to carry out.

One interesting thing to notice is the form of the *baulks*, in which connection I would like to remark that when I use the word without inverted commas I mean it in the ordinary sense as implying a big heavy timber, but when I use the commas I mean it in its technical sense as it is used in military engineering. In this latter sense it describes the timbers specially provided for the purposes just described. Large supplies of the ordinary heavy *baulks* could not be carried with an army: but strength is required nevertheless. Hence the military engineers have invented a form which combines strength with lightness.

AN INCIDENT AT LOOS
THIS PICTURE GIVES US SOME LITTLE IDEA OF THE DEVASTATION CAUSED BY MODERN
WEAPONS. IT ALSO SHOWS THE INVENTIVENESS OF THE SOLDIER WHO MAKES HIS
RIFLE INTO A BATTERING-RAM. INCIDENTALLY WE SEE A KIND-HEARTED SOLDIER
RESCUING A LITTLE GIRL FROM DANGER. THIS INCIDENT REALLY HAPPENED.

If you stand a plank upon its edge, supported at each end so as to form a beam, its strength will vary as its width and as the *square of its height*. If then you double its width you only double its strength, but if you double its height you multiply its strength *four* times. If you halve the width of a given beam you halve its strength, but if you then double its height you quadruple that half, in other words, without making the beam any heavier by these two operations you double its strength. Moreover, if you support a beam at each end and pass a load over it or spread a load permanently upon it, its greatest strength is required in the middle. You can shave away the ends without making the beam as a whole any less strong. So these *baulks* are made like planks, very oblong if looked at endwise, also thinner at the ends than in the middle. But if by chance they tipped over on to their sides they would for that very reason be very weak, and that is why the notches are provided in the transomes and the centre beams of the pontoons, in order that the *baulks*, having been laid edgewise in them, cannot tip over. Thus a considerable saving is made in the weight of the bridging material to be carried.

It sometimes happens that when a trestle is dropped into the water one leg will fall into a depression in the river-bed or will sink more deeply if the bed be soft, leaving the whole structure lop-sided and useless. That, however, is easily overcome, since it is provided against. A little iron bracket, which is carried for the purpose, is clipped on to the leg which has sunk near its top and on to it is hung a pair of pulley blocks—one of those little contrivances which everyone has seen at some time or another by which one man pulling a chain quickly can raise, although slowly, a heavy load. By this means the end of the transome is raised until it is horizontal and the legs have assumed an upright posture, when the transome is refastened to the leg in its new position. Thus we see the advantage of clamping the transome to the leg rather than fixing it with any arrangement of holes. The iron band, which is fastened on to the transome and which grasps the leg, is so arranged that the greater the load the more tightly does it hold, so that it is perfectly safe under all conditions.

The trestle bridge has a great advantage over the floating bridge if the height of the water varies at all, as for instance, with the tide. The former remains still, while the latter goes up and down, requiring a special arrangement to be contrived for connecting it to the shore.

Under some conditions a suspension bridge is the most convenient form of all, particularly if the banks are high and strong, or if the current be very rapid or the river-bed very soft. In such cases steel wire ropes are stretched across the water between two trestles. The latter may be made in the way just described, but more often they have to be stronger and are built specially out of big strong timbers securely fastened together. Their form does not matter much so long as they are strong and stiff, high enough to carry the ends of the suspension ropes and of such a shape as not to block the entrance to the bridge itself. The higher they are the better, because, according to the natural laws which govern such things, the more sag or dip there is in the ropes across the river the less severely will they be strained. They need to be very strong, as the whole weight of the bridge and its load falls upon their shoulders. The pull of the suspension ropes, moreover, tends to pull them forward into the water, so they must be held back by other strong ropes called guys, and the action of these two sets of ropes entails the unfortunate trestles bearing really *more* weight than the actual weight of the bridge and load. The guys, too, require very strong anchorage or at the critical moment they may give way, when the whole contrivance, with possibly valuable guns or ammunition on board, will be precipitated into the water. The men may be able to swim but the guns will sink.

Having, then, constructed a trestle upon each bank, securely guyed it back and connected the suspension ropes to it, the next operation is to attach smaller vertical ropes to the suspension ropes at intervals, to support the ends of the transomes. Then upon the latter are laid *baulks* and upon them the flooring as usual. Or if ropes be not sufficiently plentiful, timbers may be lashed on to the suspension ropes instead, the transomes being fastened to them.

That is all that is absolutely essential to a suspension bridge, but one so formed would be rather flimsy and unstable. It needs to be stiffened by diagonal timbers at suitable places and often it has props placed upon the bank reaching out as far as their length will permit over the water to steady and consolidate what to commence with is rather too much like a spider's web. Those little strengthening dodges can be laid down in no books. They need to be left to the judgment of the men in charge to do what is necessary in the best way they can with the materials which happen to be at hand.

But very often warfare has to be carried on in the most outlandish places where armies can only travel light, and where, hampered by bridging material of the conventional sort, they would have no chance in catching up with a fleet and agile native enemy. Yet bridges are needed even more under those conditions perhaps than under any other. There are many examples of this in the wars just beyond the frontier in Northern India. Then ingenuity has to make good the luck of prepared material and the bridges are made of those materials which happen to be procurable.

An army in India once wanted to cross a river, where no materials of the ordinary kind were available. The river, however, was lined with tall reeds. A reed has for centuries been a favourite example of weakness and untrustworthiness, so how can reeds be made to form a safe bridge? This is how it was done.

Great quantities of reeds were cut and were made up into neat round bundles about a foot in diameter. Ropes were scarce too, but these likewise were improvised by twisting long grasses into ropes. It is surprising what good ones can be made in this way, and they served their purpose well. Many bundles having thus been made numbers of them were tied together so as to form rafts. Each bundle in fact was a small pontoon, and the rafts which were thus constituted differed only in size from the regulation rafts made of pontoons.

While this work was being done two ropes were got across the river and secured on both banks: then rafts were floated down in succession, each one on arrival being tied up under the two ropes. Finally a track of boards was laid over the centre and the bridge was strong enough for men in fours to walk over it.

Had it been necessary, the floor could have been made of brushwood, interlaced so as to form a kind of continuous matting or of a layer of branches covered with canvas. Floors for bridges can be made in many ways.

A dodge which soldiers in the British Army are taught is how to make boats for bridging purposes out of a tarpaulin or piece of canvas, supported on a framework of light wood poles or twigs. The outline of the boat is first drawn roughly on the ground. Then three posts are driven in on the centre line of the boat and to the top of these three a horizontal pole is tied, thin, flexible branches stripped of their bark, being fixed by having their ends stuck in the ground on either side. The ends are driven in on the outline already marked out so that when done the branches form a framework like the ribs of a boat upside down. Other branches are intertwined among these so as to bind them together and finally a tarpaulin or canvas sheet is laid over all. A number of boats formed after this fashion can be used as pontoons to support a bridge, or several can be made into a raft and towed to and fro—a sort of floating bridge.

Another scheme is to make a number of crates like those in which crockery and other things are often packed. These are of very simple and easy construction, consisting of sticks slightly pointed at the ends driven into other pieces which are perforated with suitable holes to receive the ends. The only tools necessary are an axe (or even a pocket-knife will do) to sharpen the ends and an auger to make the holes. Almost any sort of wood can be made to serve. The cover for this, and indeed for most of these improvised rafts, is tarpaulin or canvas, the latter of which, being the material used for so many purposes, is almost sure to be available in some form or other.

For instance, every one of those familiar General Service Wagons has its large canvas cover. In fact, a general service wagon, taken off its wheels and wrapped up in its own canvas cover, makes quite a serviceable boat, pontoon, punt, barge or whatever you like to call it.

Then there is an ingenious type of little bridge which can be quickly and easily made where bamboos or similar light canes or sticks are available. The only tool required in making this is a couple of poles ten feet or so in length. To commence with, these poles are laid side by side upon the bank with one end of each pointed out over the water, overhanging it by about four feet. Two men then climb along these, while others sit upon the inshore ends to keep them from tipping into the water.

Seated, then, on the outer ends of the poles the men drive some bamboos or whatever they are using into the water, after which they tie a crosspiece to the uprights, so forming a light trestle. Then the poles are pushed forward until they overhang another four feet beyond the trestle just made, the other men, of course, continuing to sit upon the rear ends. And so the bridge grows until it entirely crosses the stream.

Between the trestles other light poles are laid and tied, forming the floor upon which men can cross in single file.

Another type, known as the *hop pole* bridge is made of slightly heavier poles which are tied together in threes so as to form isosceles triangles. Each triangle forms one trestle.

The two poles which form the sides project a little above the apex so that in fact we have an isosceles triangle with a V at the apex. To the root of the V another pole is tied loosely and the whole trestle is pushed feet first into the water. Then, by pushing the pole, it is forced into an upright position in which it is secured by the pole being firmly fixed to the shore and strongly lashed to the root of the V where, before, it was only loosely tied. A second trestle is then in like manner fixed in front of the first one, connected to it by a pole just as the first is connected to the bank. And so the thing grows. To all the upper ends of the V's a light pole is tied to form a handrail. In this case, of course, the floor of the bridge is nothing more than a pole, but with the assistance of a handrail it is quite easy to walk along a single pole.

And that reminds me of a simple type of suspension bridge which, an engineer officer once assured me, is actually copied from one habitually made by some of the Indian natives. It consists of three ropes upon one of which you walk, while the other two form a handrail upon either side. The three ropes are held at intervals in their correct relative positions by little wooden frames formed of three sticks tied together, one rope being tied to each corner of each triangle.

On the banks stakes are driven in and tied back with cords to give additional strength, and to them the ends of the ropes are secured. One drawback to this form of bridge is that the ropes are naturally far from level and one has to walk down a steep hill to commence with and up again at the other end. I once saw a specimen of this kind of bridge across a wide ditch, a part of the old defences of Chatham, and an elderly gentleman who was with me, a man of considerable proportions, insisted upon trying it for himself. He took but a step or two when his foot began to slide downhill along the foot rope faster than he dare move his hands along the hand ropes, with the result that he was very soon in a very uncomfortable position. Thus he remained, to the amusement of all his friends, until two stalwart Royal Engineers came to his aid and *uprighted* him.

In crossing a swamp something in the nature of a bridge is sometimes required. Canvas laid upon branches often makes a good road over what would otherwise be impassable.

Rapidly moving detachments of cavalry are provided with what is called *air-raft* equipment, which enables them to get their light Horse Artillery guns across rivers which would be impassable otherwise. It consists of sixty bags like huge cylindrical footballs except that the outer covering is canvas instead of leather. These are blown up partly by the mouth and partly by pumps provided for the purpose until they are just about as tight as a football should be. Then they are laid out in rows of twelve, each row being fastened together by the bags being tied to a pole running lengthwise of the row. Cords are attached to the bags for the purpose. The five rows are then placed parallel and connected together by two light planks called wheelways placed across the rows and tied thereto.

This arrangement is capable of carrying light guns or ammunition wagons. The men are expected to ride through the water, but if necessary something can be laid upon the raft, between the wheelways, to form a floor upon which men and even horses can ride.

As part of the equipment there is a small collapsible boat with oars and by its means men first cross, carrying with them a line by which, afterwards, the raft can be hauled to and fro.

Rafts can be made, too, of hay tightly tied up in waterproof groundsheets or tarpaulins or canvas. Indeed, given a little ingenuity and the need to use it (for it is very true that necessity is the mother of invention), it is surprising what a large variety of things can be pressed into this service.

Of course, barrels can be made to form excellent pontoons, but there is one clever little way of using them which is more than usually interesting, and with that I must conclude this chapter which has already exceeded its appointed limits.

Imagine two poles perhaps ten feet long, placed parallel. Between them, at one end, a barrel is lashed: at the other end is a plank forming with the poles a T. A man can then sit upon the barrel and paddle about, for the poles and planks will steady the barrel just as the outriggers and floats steady the narrow canoes or catamarans of which we read in books of travel. For that reason a bridge formed of such is called a *catamaran* bridge. Of course, if there are only a few barrels to be had they can be fitted out like this and then combined into a raft. Or if there are enough of them they can be anchored at intervals and poles or planks laid from one to another so as to form a continuous bridge. Or a single one may be used as a boat. I can almost fancy I see some of my readers who have access to a pond rigging up an old barrel in this way, just to see how it goes.

CHAPTER 7

What Guns Are Made Of

No longer ago than the days of the Crimea, the largest guns were made of the cheapest and commonest kind of iron, that known as cast iron.

This material has the advantage of being cheap and easily worked, but is comparatively weak and liable to crack, so that the guns of that time were comparatively small compared with those of to-day; they could only withstand a feeble explosion and their range was therefore limited. Had the energetic explosives of the present time been employed in them they would inevitably have burst, killing their gunners instead of the enemy. Attempts were made to strengthen them with bands made of wrought iron, a form of the metal which is tough and elastic and therefore better able to withstand sudden shocks than the more brittle cast iron, but it was not a real success.

At first sight one naturally wonders why the whole gun was not made of the stronger wrought iron. The reason was that while cast iron can be melted and poured in a liquid form into a mould, so as to produce the shape of the gun, wrought iron will not melt. It will soften with heat, in which condition it can be hammered into shape and, moreover, when in a very soft state two pieces can be joined by simply forcing them closely together, which operation is called welding.

With the machinery available now it would be possible to make a gun of wrought iron, but even a few years ago it would have been quite impossible. There was an obvious need therefore of a metal which could be melted and cast in moulds like cast iron, yet tough and strong to resist shock like wrought iron. Fortunately this problem excited the interest of a certain Mr. Henry Bessemer, a gentleman who, having made a considerable fortune through an ingenious method of manufacturing bronze powder, had sufficient leisure and money to devote himself to its solution.

The vast steel industries of Great Britain and the United States are the direct results of this gentleman's labours, and in the latter country there are quite a number of towns which, being the home of steel-works, are called by his name.

Iron is one of the most plentiful things in the world. Deposits running into millions of tons are to be found in many parts, but it is practically always in the form of ore, that is to say, in combination with something else generally oxygen and sometimes oxygen and carbon. The former sort of ore is called oxide of iron and the latter carbonate of iron, and both of them bear not the slightest resemblance to the metal. They are just rocks which form part of the earth's crust, and it is only the metallurgist who can tell what they consist of.

In order that the iron may be obtained from the ore it is necessary for the oxygen to be separated from it, an operation which requires the intervention of heat, and the heat must be obtained from a fuel which consists mainly of carbon. Wood fulfils these requirements, but there is not enough wood in the whole world to smelt all the iron which we need. It was not until *pit-cole* displaced *char-cole* (to use the spelling of the period) that the iron industry began to assume its present importance.

To produce iron cheaply, therefore, ore and coal should for preference lie side by side, and in some few favoured localities that state of things exists. Generally speaking, however, the ore and the coal are not found together, with the result that one has to be taken to the other, and in practice it is usually the ore which is taken to the coal. Hence, the iron and steelworks are generally to be found on the coalfields, while the ore comes by rail or ship from, it may be, remote parts of the world.

The method by which the metal is obtained from the ore is in principle very simple. Coal and ore are mixed together in a furnace, the fire being fanned by a powerful blast of air. The result is that the bonds uniting iron and oxygen are relaxed by the heat, when the oxygen, having a preference for union with carbon rather than with iron, leaves the latter to join up with some of the carbon of the coal.

The furnace in which this operation is carried out is a tall, vertical cylinder of iron, lined with firebrick. The fire is at the bottom and the fresh fuel and ore are thrown in at the top. As the ore is *reduced* (the chemist's term for removing oxygen from anything) the liquid iron accumulates in the lowest part of the furnace, whence it is drawn off at intervals, being allowed to run into grooves or gutters in a bed of sand, where it solidifies into what is called *pig* iron.

Along with the coal and ore, there is thrown into the furnace from time to time quantities of limestone which combines with the earthy impurities with which the ore is contaminated. Together these form what is called *slag*, which also exists, while in the furnace, as a liquid, but is so much lighter than the molten iron that it keeps quite separate and can periodically be drawn off through a hole higher up than that through which the iron is obtained. The slag solidifies into a hard stone which is broken up and used for making concrete and tar-paving, also for road metal.

The kind of furnace just described is, owing to the strong blast of air needed for its operation, called a *blast-furnace*. One would be inclined to think that a fire so well supplied with oxygen, both from the blast and from the ore itself, would cause the fuel to be completely burnt up, yet such is not the case. The gases which ascend from the fire consist largely of *carbon-monoxide*, a burnable gas with lots of heat still left in it. Years ago, and one may still see instances of it, this gas was allowed to escape at the top of the furnace, where it burnt in the form of a huge flame. In most modern furnaces, however, there is a kind of plug in the orifice at the top which, while it can be lowered in order to admit the ore and fuel, normally prevents the escape of the gases, which are led away through pipes. In some cases the gases are burnt under boilers to provide the works with steam, in other cases they heat other furnaces for metallurgical purposes, while in yet others they are employed to drive large gas-engines to generate electricity. It is sometimes a difficulty to find useful employment for the vast quantities of this blast-furnace gas which are produced at a large works.

We see, then, how is obtained the pig iron from which the other kinds of iron and steel are made. It is not pure iron by any means; indeed, it is not sought to make iron pure, as is the case with most other metals, since, in its pure state, it is too soft to be of much use. All the familiar forms of iron and steel are really alloys of iron and carbon, a fact which tends to give iron its unique position among the metals, since by exceedingly slight variations in the percentage of carbon we can vary the properties of the iron to an amazing extent, thereby producing in effect a wide range of different substances each particularly suitable for a particular purpose.

To make cast iron, such as the guns of the Crimea were made of, it is only necessary to melt up some pig iron and to pour it into a mould. There is scarcely a town in which there is not an iron foundry, either large or small, and that is the work carried on there. A smaller form of the blast-furnace, known as a *cupola*, melts the pig iron, and the moulds

An 18-pounder in action

The crew consists of six men. No. 1 (the sergeant) gives instructions. No. 2 stands at the right of the breech. No. 3 fires the gun. No. 4 holds the shell ready for placing in the bore. No. 5 adjusts the fuse and hands the shell to No. 4. No. 6 prepares the ammunition and hands it to No. 5. In this picture only three of the crew are left.

are generally made of sand. The process of pouring the melted metal into the moulds is called *casting* and the things so produced are *castings*, and are said to be made of *cast* iron.

Wrought iron is made by working the molten pig iron instead of casting it. The work is done in a different type of furnace altogether from the blast furnace and the cupola. It is more like an oven, in the floor of which is a depression wherein the molten metal lies. The fireplace is so arranged that the flames pass over the metal, being deflected downwards upon it by the roof as they pass.

It should be understood that in casting pig iron one does little more than form it into some desired shape, the nature of the metal undergoing little or no change. In working it, however, into wrought iron, we change its nature.

The pig iron contains from 2 to 5 per cent of carbon, which it obtains from the coal in the blast-furnace, and it is this particular proportion of carbon which gives it its own peculiar properties. To convert it into wrought iron a workman puts a long iron rod into the furnace and stirs the metal about, thereby exposing it to the air and permitting the carbon to be burnt out. As it loses carbon the iron becomes less and less fluid until it reaches a sticky stage. Thus the workman, who is known by the name of puddler, as the process is called puddling, works up a ball of decarbonized and therefore sticky iron upon the end of his rod. Having thus produced a rough ball or lump he draws it out of the furnace and leaves it to cool.

Thus the result of the puddling process is to produce a number of rough lumps or balls of iron with only about one-tenth per cent of carbon. They are next reheated, in another furnace, and a number of them are hammered together under a mechanical hammer into larger lumps called blooms or billets. The hammering process has the effect of driving out impurities and also of improving the texture of the metal.

Iron sheets, bars, rods and so on are formed by heating the billets and rolling them out in powerful rolling mills, machines which in principle are precisely similar to the domestic mangle, wherein two iron rollers with properly shaped grooves in them squeeze out the billet into the desired form.

Wrought iron, owing to the method by which it is produced, is not homogeneous, that is to say, it is nor quite the same all through, with the result that when it is rolled it develops a grain somewhat similar to the grain in wood, so that if bent across the grain it is somewhat liable to crack. On the other hand, it has the advantage over steel that it rusts much less readily. Hence, for outdoor purposes it is still sometimes preferred to the otherwise more popular steel.

Now the problem which Bessemer set before himself was to find out how to make a metal which could be cast like cast iron yet should be as strong and tough as wrought iron. After a little experimenting, by a happy inspiration, he hit upon the idea of blowing air through a mass of molten pig iron, thereby burning out the carbon, just as is done in the puddling process, only much quicker and with less labour. By this means he produced a metal with less carbon than cast iron and more than wrought iron, a sort of intermediate state between the two, and to his joy he found that this Bessemer steel could be cast like cast iron yet had strength and toughness equal to if not superior to that of wrought iron. Moreover, it was homogeneous and when rolled did not possess the troublesome grain characteristic of wrought iron.

Having thus found the way to make this new and desirable metal, Bessemer encountered a great disappointment, so great that it would have entirely beaten many men. He made samples of steel and submitted them to experts in iron manufacture. Everyone thought them admirable and many large iron works were induced by them to make arrangements with Bessemer for the right to use his process. His name was already famous and it seemed as if a new fortune was made, when, to his alarm, he learned that wherever it was tried except in his own works, the process was a miserable failure. Instead of being at the end of his labours he was just at the beginning.

It turned out that the particular iron which he happened to buy and use at his own works was particularly free from an impurity which is, generally speaking, a great nuisance in iron, namely, phosphorus. It was pure accident which had led him to use this iron: it happened to be the kind he could purchase most easily in the small quantities needed for his experiments but it led him into a great difficulty, for other people, after paying him for the right to use his process and after spending large sums on the requisite plant, found themselves unable to make the steel because of the phosphorus in their iron and finding themselves unable to make a success were inclined to write him down a fraud. As it turned out, after much labour on Bessemer's part, it was due to the presence of tiny percentages of phosphorus in most of the iron that is produced.

After much trouble he was able to induce certain owners of blast-furnaces to make, by special methods, a kind of pig iron practically free from phosphorus and therefore suitable for his process. This special pig iron was known as Bessemer Pig Iron.

A little later a new inventor, a Welshman, Thomas by name, over-came the difficulty in another way, but to explain that I must first describe the Bessemer Converter, the special apparatus designed by Bessemer for making his steel.

It can best be likened to a huge iron kettle with a big spout at the top and with two projecting pins, one on each side. These pins rest in supports, so that it is easy to tilt the whole thing over on to its side. This is lined with fire-clay or some suitable heat-resisting material.

Through one of the *pins* (*trunnions* is their proper name) there runs a hole, communicating to what we might call a grating in the bottom of the converter. To this hollow *trunnion* there is connected the pipe from a powerful blowing engine, so that air can be driven in at will.

To load or charge the converter it is tilted over somewhat to one side so that molten pig iron can be poured into it. The blast is then turned on after which it is raised to an upright position with the air bubbling up from below through the iron. Thus by being brought into close contact with air, the carbon is burnt out of the metal until none is left. That, however, is not desired, so, as soon as the carbon is known to have all gone, a fresh quantity of molten iron is added of a special kind, the amount of carbon in which is known very exactly. Thus all the carbon is first removed and then exactly the right amount is added, and so the desired result is attained with certainty.

Now Thomas's improvement was this. He discovered that the converter could be lined with certain substances which have a great attraction for phosphorus and under those conditions any phosphorus which may be in the ore goes readily from the iron into the lining, or forms, with material from the lining, a slag which floats upon the surface of the metal.

When the process is completed the converter is tipped over once more and the metal, now steel, is poured into rectangular moulds from which the steel can be lifted after cooling in the form of ingots.

Steel produced by Bessemer's process as improved by Thomas is called Basic Bessemer Steel.

Incidentally Thomas, by this invention, laid the foundation of much of the steel industry of Germany and Belgium, for there are enormous deposits of ore in the neighbourhood of Luxemburg which because of the presence of phosphorus were useless until Thomas showed how it could be dealt with.

And there is another interesting feature of this basic process. Phosphorus is a valuable fertilizer, so that the *slag* makes a very fine chemical manure. It is ground up into a fine powder and is sold to farmers under the name of Thomas's Phosphate Powder. It owes its fertilizing virtues to the presence of the phosphorus which it has stolen from the molten iron.

Bessemer derived a huge fortune from his process after he had fought and overcome his difficulties, in addition to which he received the honour of knighthood and became Sir Henry Bessemer.

It will be noticed that one of the virtues of the process is its economy in fuel. During the whole time that the metal is in the converter, from twenty to thirty minutes, no fuel is used to keep it hot. The reason for that is that the carbon which is being got rid of is acting as fuel. It is burning with the air which is driven through, thereby generating heat.

In Bessemer's early days, it was arranged that he should attend a meeting of ironmasters at Birmingham to explain his new process. On the morning of his lecture two eminent ironmasters were breakfasting together in a Birmingham hotel when one exclaimed to the other, *What do you think, there is a fellow coming here to-day to tell us how to make steel without fuel.* To this eminent South Wales ironmaster the proposal seemed preposterous but it was true all the same.

Although vast quantities of steel are made by the Bessemer process there is another one of equal importance known as the Siemens-Martin Open-hearth process. In this the molten metal is kept in a huge bath practically boiling until the carbon has been reduced to the required amount. Perhaps the most interesting feature about it is the way in which fuel is saved by what is called the *regenerative* method due to that versatile genius Sir William Siemens.

The open-hearth, as it is termed, is a huge rectangular chamber of firebrick with a firebrick roof, and doors along one side just under the roof through which the process can be watched and new materials be added from time to time.

The fire is some way away and not underneath as one might perhaps expect. Now if a deep coke fire is fed with insufficient air it does not give off carbonic acid such as usually arises from a fire, and which as everyone knows will not burn, but a gas called carbon monoxide which will burn very well. So the fire-place for these furnaces is constructed in such a manner as to produce carbon monoxide, which then passes through a huge flue to one end of the open-hearth. Here it meets air coming through another flue and the two combining burst into flame over the metal.

The hot gases resulting from this burning pass out through a flue at the other end of the hearth to a tall chimney which causes the necessary draught, but on their way they pass through a chamber loosely filled with bricks. Consequently the hot gases only reach the open air after having given up much of their heat to these bricks.

After that operation has been going on for a time certain valves are operated and the gas and air then come in at the other end of the hearth, travelling through it in the opposite direction. And the air comes through the chamber which has the hot bricks in it, bringing back into the furnace a large quantity of that heat which otherwise would have gone up the chimney but which the bricks intercepted. Thus all day long does this reversal take place at intervals, the fresh air all the time picking up and bringing back some of the heat which just previously had escaped towards but not into the chimney. This arrangement enables the process to compete, so far as economy is concerned, with the Bessemer process.

At intervals the steel is tapped off from the furnace and run into ingot-moulds, the same as with the other process. On the whole it is regarded as producing a slightly better steel, the operation being under slightly better control.

However the steel is made the ingots are reheated and either hammered under a powerful steam hammer or pressed in an enormous hydraulic press. This greatly improves the quality.

The steel can then be rolled into plates, bars or whatever form may be required.

The finer qualities of steel such as are used for making sharp tools are made in quite another way. Instead of being made from crude iron by taking out the carbon, the materials are the finest qualities of wrought iron and charcoal which are mixed together in the correct quantities and melted in a crucible. This cast steel is very hard, so that it will carry a very fine, sharp edge. It is also capable of being tempered by heating and cooling, so that the exact degrees of hardness and toughness can be attained.

Of recent years a special quality of steel for tools called high-speed steel has been produced, mainly by the addition to ordinary cast steel of a small percentage of tungsten. The advantage of this is that, within certain limits, this does not soften with heat, and it is, I can assure you, a great invention in war-time, when a nation is straining every nerve to turn out guns and shells as fast as possible.

For all these things need to be turned in lathes and if you have ever watched a metal-turning lathe at work you will have noticed that the tool which actually takes a shaving off the article being turned tends to get hot. For this reason lathes are usually fitted with pumps which pump cold soap-suds on to the tool as it works. What you see there is the energy employed in shaving the metal being turned into heat in the tool. If left uncooled by the water it would soon be red-hot. And the faster the machine works the hotter will the tool get.

Now with the old steel a very little heat will suffice to make it soft, when its cutting power is lost. So with the old steel, no matter how much cooling water you might use, there was a distinct limit to the speed of the lathe and the speed at which the work was finished, for if that speed were once exceeded a stop became necessary to regrind the tool or to put in a fresh one.

But with high-speed steel that limit is much higher, for it can get almost red-hot before it loses its hardness and consequently machines can be run and jobs finished at a speed which would have been out of the question only a few years ago. If one belligerent knew how to make high-speed steel while the other did not the former would have an enormous advantage in war-time.

Speaking generally, steel such as is used for tools is called hard steel, while that made by the Bessemer and Siemens–Martin processes is called mild steel. Leaving out of account for the moment fancy steels such as that just described, where other metals are added to the mixture, the essential difference between all the varieties of steel is simply a slight difference in the percentage of carbon. This is so remarkable that it is worth while to tabulate these percentages again.

Cast iron has from 2 to 5 per cent.

Steel from one-fifth to one per cent.

Wrought iron less than one-fifth per cent.

Mild steel, which has least carbon of all the varieties of steel and in this respect is therefore nearest to wrought iron, is used for the same purposes as wrought iron, such as shipbuilding, bridges and roofs, tanks, gas-holders, etc. When the Admiralty want a specially fast ship such as a torpedo-boat destroyer with a hull as light as possible consistent with strength they have it made of steel with a slightly larger percentage of carbon so that the steel is stronger and the vessel's frame can be made lighter. The steel for shells, too, needs to be of a certain strength to give the best results, so the percentage of carbon is adjusted accordingly.

For guns themselves, again, special properties are needed, and so not only is the carbon regulated to a nicety but other things such as nickel and chromium are added. Altogether, steel is one of the most marvellous substances known, certainly the most marvellous metal. Copper is just copper and no more, zinc is just zinc, and the same with lead, but iron (which really includes steel) can be adapted to so many purposes, can be endowed at will with so many different properties, that without doubt iron, common, plentiful iron, is the king of all the metals.

CHAPTER 8

More About Guns

As has been remarked elsewhere, some of the guns used by the soldiers in land warfare are very different from those used in the navy. The latter, being carried on the ships to which they belong, can be of those proportions which best suit their purpose. Consequently they are usually very long compared with their diameter.

The field guns used by the Royal Field Artillery are shorter in proportion to their calibre than are the big naval guns. Otherwise they would be far too long to handle in the field. They are mounted on carriages drawn by horses, and are so handy that they can go anywhere where infantry can go and can travel just as fast. It takes a very short time to get them ready for action, too, so that they can accompany infantry quite freely, neither arm impeding the movements of the other. The Horse Artillery, again, whose guns are even lighter still, can accompany cavalry, travelling as fast and coming into action almost as quickly as the troopers themselves.

The famous French *seventy-fives* (meaning 75 millimetres calibre) which played such a great part in the war, are field guns intended to move rapidly and to operate with infantry.

Both these types of gun were used by the British in South Africa, as also were some field howitzers, a type of gun to which further reference will be made later. But the Boers taught the world something new as to the possibilities of moving heavy guns quickly. Perhaps the reason for this was that they, being something of the nature of amateurs in the art of warfare, were less under the influence of tradition. Anyway, they surprised the British by the quick way in which they moved heavy guns, sometimes into quite difficult positions, over rough ground and up steep hills. These heavy guns of theirs were called by the British soldiers Long Toms.

But the British were quick to respond, particularly the ever-resourceful navy. When the war broke out there were, in the neighbourhood of Durban, a number of warships which had as part of their own armament some of those guns which afterwards became famous as 4·7's, that being the diameter of the bore in inches. They were of the long shape usual in naval guns, and it is easy to see that they were much heavier than the field guns of 3 inches or so in diameter.

Captain Scott (now Admiral Sir Percy Scott) saw that these would be useful, so he quickly designed some carriages for them, got these made in the railway workshops at Durban, and in a few hours was rushing them up to Ladysmith. It was these guns very largely which enabled that town to hold out for so long, until, in fact, it was triumphantly relieved.

Thus the effect of the Boer war was to show that much heavier weapons could be manipulated in the field than had been considered possible before. The Great War which followed but a few years later carried on this same lesson, for one of the great surprises with which the Allies were confronted in the early days of the conflict was the inexplicable fall of fortresses which till then had been deemed almost impregnable.

Liége, Namur, Maubeuge and, finally, Antwerp, all fell to a wonderful gun of enormous dimensions which the Austrians had produced from up their sleeve, so to speak. Like conjurers they had kept them secret until the last moment.

These weapons which made history so fast were of the kind called howitzers, a name mentioned just now. It should be explained here that gunners talk of guns and howitzers as if the latter were not guns; but that is only a convenient habit which has grown up, for the latter are unquestionably guns. The distinction is, however, so convenient that we may well adopt it ourselves for the rest of this chapter.

Repeated references have been made already to the question of the length of guns, and it has been pointed out that to get high velocity, great range and vigorous hitting power a gun needs to be as long as possible. On ships this is only limited by the strength of the steel of which the gun is made, for beyond a certain length the gun bends of its own weight. Ashore, however, the difficulties of transport impose a further limitation in most cases, although the famous 4·7, like many other naval guns, has a length of 50 calibres, and the guns of small calibre do approximate somewhat to the proportions of the naval guns, since even then their length comes within manageable limits.

Modern warfare, however, requires the use of larger shells containing larger charges of explosives, and to fire these requires guns of greater calibre. We hear of shells of as great a diameter as 16 inches being thrown into the Belgian fortresses and of course nothing smaller than a 16-inch gun could do that. Now a 16-inch gun, if made to the naval proportions of 50 calibres or even 45 calibres, would mean a length of at least 60 to 70 feet. It would also mean a weight exceeding 100 tons, for the 12-inch naval gun of 50 calibres weighs about 70 tons. And it is easy to see that such a gun would be very difficult to move on the field of battle. Indeed, it would be almost useless because of the time it would take to get it into position and to construct the foundations which it would need. If the Austrians had only had such as those the Belgians would have had plenty of time to prepare for them at Antwerp, whereas it was the quickness with which they brought up their heavy guns that astonished everyone and took their opponents by surprise.

The secret of this astonishing performance lies in the fact that they were not guns at all but howitzers, which instead of being long, slender tubes are short, fat ones, and that involves a different idea in gunnery altogether. The gun fires *at* an object. The howitzer fires its shell upwards with the purpose of dropping it *upon* the object.

The difference between the two is well illustrated by the methods of practising with them. In learning to work a gun the gunners fire at a vertical target just as those of you who practise shooting at a miniature range fire at a target of paper placed vertically against a wall. The target for howitzer practice, on the other hand, is a square marked out on the level ground, and the object of the gunners is to see how great a proportion of a given number of shots they can drop inside that square.

Of course, being so much shorter the howitzers cannot throw a shell so far or at such a high velocity as the naval guns, but that can to a certain extent be compensated for by using a higher explosive for the propellant. That, however, involves greater stresses in the tube when firing takes place and also calls for stronger foundations in order that the aim may be steady.

A great part, too, of the velocity of a naval shell is required for the penetration of the armour, whereas against forts or earthworks it is sufficient if the shell *gets there*.

Moreover, generally speaking, it is possible to get much nearer to a fortress or entrenched position for the purpose of attacking it than it is to an enemy ship on the sea. Except for the occasional help of a mist there is no cover to be obtained at sea, while on land the ground must be very flat indeed if there is no low hill or undulation behind which a gun can be set up unnoticed.

A GERMAN AUTOMATIC PISTOL

THE ACTION IS FULLY DESCRIBED ON THE ILLUSTRATION.

The Austrians cherish a piece of steelwork from one of the forts of Antwerp which they smashed with a shell from one of their big howitzers at a range of seven miles. They evidently were able to get their big howitzers within that comparatively short distance of the Antwerp fortifications without being molested.

In this connection one often hears the word mortar used, and just a reference to that will be appropriate here. Many years ago short guns which threw their balls very high were in use, and because of their resemblance to the mortar which is used for pounding up things with the aid of a pestle these were termed mortars. Later a man named Howitzer introduced a type of gun which was something of a compromise between the long thin gun and the short stubby mortar. As time has gone on, however, the mortars have grown in length while the howitzers have shortened, until to-day the two names are used almost indiscriminately to denote the same thing. Hence the giant howitzers of the Austrians are often spoken of as the *Skoda* mortars, Skoda being the name of the factory where they were made.

At one time many people wondered why the Germans did not put some of these huge mortars on their battleships: many thought that they would do so, and that by that means they would demolish our navy as they had already smashed the Belgian forts. The reason they did not is, no doubt, the very simple one, that our naval guns would have probably sunk their ships before the howitzers could have reached ours, because if they had attempted to make up for the shortness of the weapons by using higher explosives, these mortars would, there is little doubt, have knocked to pieces the ships on which they were mounted.

The old-fashioned fortress, suddenly made out-of-date by the Skoda mortars, was usually armed with guns of the naval type. Sea-coast forts are always so armed. Nowadays, however, the inland fortress takes the form of a labyrinth of trenches and underground passages, combined with deeply excavated chambers known as dug-outs, and these do not fitly accommodate large guns at all. The guns are placed well back behind the trenches sheltered behind hills or woods, over which they hurl their shells. The chief defenders of the actual trench are the machine gun, which is little more than an automatic rifle on a stand, and the trench mortar.

We are now in a position to sum up broadly the features of modern artillery. There is first the naval gun, the ideal gun, long and of great range, able to send forth its shells with great velocity. This gun appears again in the sea-coast forts, where the conditions are very much those which obtain on a ship and where the attacking party is of necessity a ship.

In the field we have the field and horse artillery, which we may regard as the naval gun modified somewhat in order to make it easy to move about, so that it can accompany troops and support the operations of both infantry and cavalry. These light guns are supported by the field howitzers, which are also light and easily handled, and the guns of the 4·7 type, originally naval guns but now mounted on wheels and possessing a certain amount of mobility, not equalling the field guns it is true, but still very serviceable in a campaign.

Then we have the howitzers of various sizes which have rendered the old-fashioned steel and concrete forts useless, and which are the chief weapons used in the modern trench warfare. It is these which blow in the walls of the trenches and dug-outs, shatter the barbed-wire entanglements and render it possible for the infantry to attack an entrenched position.

Finally, we have the machine guns, each of which is equivalent to a considerable number of riflemen and which, with the trench mortars, form the chief defences of the actual trench itself. Of course these are only useful against attacks by infantry: they cannot in any way cope with the heavy artillery. That has to be dealt with by the opposing artillery posted away back behind the trenches.

And now let us take a rather more close look at some of these weapons. Essentially each one is a steel tube. It may be a single tube or it may be several one outside another. It may even have a layer of wire between two tubes as many naval guns have. It is invariably (one small exception will be mentioned later) loaded at the breech or rear end and not through the muzzle as used to be the custom. For this purpose it needs a breech-block or door, which can be opened to put in the shell and explosive, and which can then be closed tightly so that it will not be driven out or burst open when the explosion takes place and also shall be gas tight so as not to let any of the force of the explosion escape.

Then the gun must be mounted upon a carriage so that it can be quickly moved about. The lighter forms of artillery are fired when upon the same carriage upon which they travel. In years gone by the whole thing, carriage as well as gun, used to run back when the gun was fired, which was a great nuisance since it had to be got back into position again after each shot. To obviate this the gun is now mounted upon a slide, and it is the slide which is fitted to the carriage. Thus the gun can slide back without the carriage moving at all. The latter is made very strong, and shoes are provided at the end of chains which go under the wheels just like the *drag* which coaches and heavy carts have for use going down hills. There is also a part like a spade which can be driven down into the ground so that, what with the shoes and the spade, the carriage is fixed very firmly.

The gun is kept at the front part of the slide by means of a powerful spring, which is compressed when the gun is fired but which, as the force of the recoil is spent, pushes the gun back to its original position once more. The spring is often reinforced by a cylinder and a piston with compressed air or water behind it, acting after the manner of those door checks with which we are all familiar, its function being to steady the motion of the gun and to let it go gently back to its place without slamming, just as the door check prevents a door from slamming.

By this means the gun is returned automatically after each shot to practically the same position which it occupied before, so that it does not need re-aiming each time, but only a slight readjustment if even that. The result of this is that such a gun can be fired very rapidly. In fact, it can be fired just as fast as the gunners can keep on reloading it.

The big Skoda mortars owed their mobility to the clever way in which they were constructed. The gun tube itself, the support for it or mounting, and the steel foundation were each fitted to a special motor-driven trolley. The steel foundation was dumped down on the ground, which of course was prepared for it in advance, then the mounting was run right on to it so that it simply needed bolting down and finally the tube was hoisted by specially prepared appliances into its place. It is said that the whole operation occupied less than an hour.

For firing, these mortars of course are pointed at a very high angle, almost like an astronomical telescope. No doubt the gunners have many jokes about *shooting the moon* and so on, for that is just what they seem to be attempting. For loading, however, they are lowered into a horizontal position: the shell comes up on a small hand-truck, is raised by a specially designed jack until it is level with the breech, and is then pushed into its place. The breech is then closed, the tube re-elevated, and all is ready for firing.

Between these two forms of gun, the field gun on its light carriage, which not only bears it from place to place but forms its support while in action, and the great mortar carried in parts on specially made trolleys, there are now an enormous variety of guns and mortars adapted for the various purposes which experience in the Great War revealed. Artillery suffered many changes in the light of the South African campaign and of the Russo-Japanese war, but of far more importance have been the lessons learnt in Northern France and on the plains of Poland. To some extent these lessons have been learnt and profited by during the actual war, but there is no doubt that as men have time to think over them in the years of peace which are ahead many more developments will take place. Unless, that is, we are on the threshold of that happy time when guns and fighting material of all sorts will be looked upon as the relics of a bad and ruinous time now happily past.

In conclusion, a passing reference must be made to the trench mortars and similar contrivances which have arisen as the result of the prolonged spell of trench warfare which no one had ever contemplated. These are in effect very short range mortars or howitzers, specially intended for throwing bombs from trench to trench. Some are simply the larger mortars on a small scale, but one has decidedly original features.

This consists of a short light mortar into which the bombs are slipped through the muzzle, thus reverting to the old method of loading. The propellant is combined with the bomb and there is a percussion cap which fires it as soon as it strikes the bottom of the tube. Thus the operation is just about as simple as it can be: the man merely places the bomb in the upturned muzzle and lets it slide down. An instant later, up it comes again, to go sailing through the air into the trench of the enemy a hundred yards away.

One must not conclude this chapter, however, without a reference to those useful weapons which are known among the soldiers as Archibalds and officially as anti-aircraft guns. These are perhaps the most familiar guns of all to the general public, since they were installed in many places in Britain for the purpose of dealing with the Zeppelins. No doubt not a few of my readers have had the experience of being awakened from their beauty sleep by the cracking of the anti-aircraft guns and have seen their shells bursting like squibs in the air.

They are fairly long guns, not unlike field guns, but they are mounted upon special supports which enable them to be pointed at any angle so that they can fire right up into the sky. The sights, also, are somewhat different, being fitted with prisms, or reflectors, so that the gunners can look along the sights and align the gun upon an object overhead without lying on their backs.

Much more could be said on this subject, but national interests forbid, so with this general review of modern artillery we must pass to another subject.

CHAPTER 9

The Guns They Use in the Navy

Both the great English-speaking nations are immensely proud of their navies. They can, on occasion, produce soldiers by the million of the very highest and most efficient type, but they never feel quite that pride and patriotic fervour over their soldiers that they do over their ships of war and their sailors.

The guns, therefore, with which the ships are armed, always form a subject of great interest, especially those large ones which constitute the armament of the Dreadnought battleships and battle-cruisers.

Let us first consider what is required in a naval gun, for it must be remembered that the naval and military weapons are different in some respects. Experience at the Dardanelles showed that even the guns of the *Queen Elizabeth*, the largest and most powerful then known, fresh from the finest factories, were not particularly successful against the Turkish forts. The Germans, too, set up what was probably a naval gun and occasionally dropped shells into Dunkirk with it at a range of twenty miles or so, but without causing much harm, and the fact that they only did it occasionally and then abandoned it altogether seems to indicate that in their opinion they were not doing much good with it.

It must not be assumed from this that naval guns are bad guns or poor guns, however, but simply that they are made for a special purpose for which they are highly efficient, from which it follows almost as a natural consequence that they are somewhat less efficient when used for some other purpose. Their purpose is to pierce the hard steel armour with which warships are protected and then to explode in the enemy's interior, whereas in modern warfare the greatest military guns are chiefly required to blow a big hole in the ground or to shatter a block of concrete. In both cases the ultimate object is to carry a quantity of explosive into the enemy's territory and there explode it, but whereas the land gun has simply to do that and no more, the naval gun has to pierce thick armour-plate as well.

And just think what that means. Many large ships have their vital parts protected by armour-plates twelve inches thick. Moreover, the armour-plates are made of very special steel, the finest that can be invented for the purpose. Vast sums of money have been expended in experimenting to find out just the best sort of steel for resisting penetration by shells. Some time ago I saw several pieces of armour-plate which had been used in one of these tests. They had been set up under conditions as nearly as possible the same as those obtaining on the side of a ship and then they had been fired at from varying distances, the effects of the various shots being carefully recorded. And that is only one experiment out of tens of thousands which have been tried again and again, while the steel manufacturers are always trying to improve and again improve the shell-resisting properties of their steel. Thus, we see, the presence of the steel armour which has to be perforated before the shell can do its work makes the task set before the naval gun somewhat different from that which confronts its military brother.

These considerations result in the naval gun needing to have as flat a trajectory as possible and its projectiles the highest possible speed.

Now trajectory, it may be useful to explain, is the technical term employed to denote the course of a projectile, which is always more or less curved.

Let us imagine that we see a gun, pointed in a perfectly horizontal direction, and let us also imagine that by some miracle we have got rid of the force of gravity and also that there is no air. Under those conditions the shot from the gun would go perfectly straight and with undiminished velocity for ever and ever. Then let us imagine that the air comes into being. The effect of that is to act as a brake which gradually slows the shell down until finally it stops it. Theoretically, perhaps, it would never quite stop it, but for all practical purposes it would.

Again, let us suppose that while the air is absent the force of gravity comes into play, what effect will that have? It will gradually pull the shell downwards out of its horizontal course, making it describe a beautiful curve.

But, someone may think, does not a rapidly-moving body remain to some extent unaffected by gravity? Not at all: it falls just the same and just as quickly as if it were falling straight down.

If our imaginary horizontal gun were set at a height of sixteen feet and a shell were just pushed out of it so that it fell straight down the shell would touch the ground in one second. If the ground were perfectly flat and the shell were fired so that it reached a point half a mile away *in one second* it would strike the ground exactly half a mile away. You see, the horizontal motion due to the explosion in the gun and the downward motion due to gravity go on simultaneously and the two combined produce the curve.

To make this quite clear, let us imagine two guns precisely alike side by side and both pointed perfectly horizontally. From one the shell is just pushed out: from the other it is fired at the highest velocity attainable: both those shells will fall sixteen feet or a shade more in one second, and if the ground were perfectly level both would strike the ground at the same moment although a great distance apart.

Clearly, then, the faster the shell is travelling the more nearly horizontally will it move, for it will have less time in which to fall, and the slower the more curved will be its path, from which we see that the air by reducing the velocity causes the curve to become steeper and steeper as the shell proceeds.

If, then, our gun is placed low down, as it must be on a ship, to get the longest range we must point it more or less upwards because otherwise the shell will fall into the water before it has reached its target. When we do that we complicate matters somewhat, for gravity tends to reduce the velocity while the shell is rising and to add to it again while it is falling. We need not go too deeply into that, however, so long as we realize that, whatever the conditions may be, the shell in actual use has to follow a curved course, first rising and then falling.

The really important part about a shell's journey is the end. So long as it hits it really does not matter what it does on the way, and if it misses it is equally immaterial. The reason why we need to bother about the first part of the trip is because upon it depends the final result. Whatever the trajectory may be we see that the shell must necessarily arrive in a slanting direction. And the more steeply slanting that direction is *the less likely is the target to be hit*.

If the shell went straight it would only be necessary to point the gun in the right direction and the object would be hit no matter how far away it might be. The more curved the course is, the more likely the shell is to fall either too near or too far, in the one case dropping into the water, in the other passing clear over the opposing ship.

Let us look at it another way. Suppose the vital parts of a ship rise 20 feet out of the water and the shell arrives at such an angle that it falls 20 feet in 100 yards: then, if the ship be within a certain zone 100 yards wide it will be hit in a vital spot. If it be nearer the shell will pass over, if it be further the shell will fall into the water. That 100 yards is what is called the danger zone. If the shell is falling less steeply, say, 20 feet in 200 yards, then the danger zone is increased to 200 yards and so on, which gives us the rule that the flatter the trajectory, or the more nearly straight the course of the shell the greater is the danger zone and the more likely is the enemy ship to be hit.

We have established two facts, therefore, first, that the trajectory must be as flat as possible and, second, that to make it flat the velocity must be high. We can also see another reason for high velocity, namely, to give penetrating power.

To obtain a high velocity the gun must be long, and consequently naval guns are always long, a fact which is very noticeable in the photographs of warships. The reason for this is quite obvious after a little thought. You could not throw a cricket ball very far if you could only move your hand through a distance of one foot. To get the best result you instinctively reach as far back as ever you can and then reach forward as far as you are able, so that the ball shall have as long a journey as possible in your hand. Perhaps you do not know it but all the time you are moving your hand with the ball in it you are putting energy into that ball, which energy carries it along after you have let go of it. And it is just the same with the shell in the gun. So long as it is in the gun energy is being added to it but as soon as it leaves the muzzle that ceases. After that it has to pursue its own way under the influence of the energy which has been imparted to it.

The powder which is employed as the propellant or driving power is of such a nature and so adjusted as to quantity that as far as possible it shall give a comparatively slow steady push rather than a sudden shock, so as to make full use of the gun's length, the expanding gases following up the shell as it goes forward and keeping a constant push upon it.

On the other hand, a gun can be too long, for no steel is infinitely strong and stiff, so that beyond a certain limit the muzzle of the gun would be likely to droop slightly of its own weight and so make the shooting inaccurate. The limit seems to be about 50 calibres or, in other words, fifty times the diameter of the bore.

For a considerable time the standard big gun of the British Navy was the 12-inch, that being the calibre or diameter of the bore. The famous *Dreadnought* had guns of that calibre and so had her immediate successors.

The 12-inch gun of fifty calibres weighs 69 tons and fires a projectile weighing 850 lbs. which it hurls from its muzzle at a velocity of about 3000 feet per second.

More recently the size has grown to 13½, 14 and even as great as 15 inches calibre, but we may for the moment take the 12-inch gun as typical of all these large guns and have a look at its construction.

It is made of a special kind of steel known as nickel-chrome gun steel, formed by adding certain proportions of the two rare metals nickel and chromium to the mixture of iron and carbon which we ordinarily call steel. The metal is made after the manner described in another chapter and is cast into the form of suitably-sized ingots which are afterwards squeezed in enormous hydraulic presses into the rough shape required. Besides giving the metal the desired form this action has the effect of improving its quality. Since a gun is necessarily a tube it may be wondered why the steel is not cast straight away into that shape instead of into a solid block and the reason why that is not done is very interesting. It is found that any impurities in the metal—and it is impossible to make it without some impurities—collect in that part which cools last and obviously that part of a block which cools last is the centre. Thus the impurities gather together in the centre of the mass whence they are removed when that centre is cut away, whereas if the first casting were a tube they would collect in a part which would re-main in the finished gun.

The ingot, then, is cast and pressed roughly to shape. Then it is put into a lathe where it is turned on the outside and a hole bored right through the centre.

But that is by no means all of the troubles through which this piece of steel has to pass. It undergoes a very stringent heat treatment, be-ing alternately heated in a furnace to some precise temperature and then plunged into oil, whereby the exact degree of hardness required is attained.

Moreover, this is only one of the tubes which go to make up the gun, which is a composite structure of four tubes placed one over another with a layer of tightly wound wire as well.

First, there is the innermost tube, the whole length of the gun, then a second one outside that, usually made in two halves. Both are carefully made to fit, and then the outer is expanded by heat to enable it to be slidden over the inner one, after which on cooling it contracts and fits tightly. Outside this second tube is wound the wire, or more strictly speaking tape, for it is a quarter of an inch wide and a sixteenth thick. It is so strong that a single strand of it could sustain a ton and a half. It is carefully wound on; first several layers running the whole length of the gun and then extra layers where the greatest stresses come, that is to say, near the breech, for that has to withstand the initial shock of the explosion. Altogether about 130 miles of wire go on a single gun.

The advantages of this form of construction are many. For one thing, a wire or strip can be examined throughout its whole length and any defect is sure to be found, whereas in a solid piece of steel, no matter how carefully it may be made, there may lurk hidden defects. Moreover, if a solid tube develops a crack anywhere it is liable to spread, whereas a few strands of wire may be broken without in any way affecting the rest. It has been found that even if a shell burst while inside one of these guns no harm is done to the men in the turret where it stands, a thing which cannot be said for guns composed entirely of tubes, so that the merit rests with the wire. A third advantage is that the wire can be wound on to the tube beneath it at precisely that tension which is calculated to give the best result, whereas in shrinking one tube on to another this cannot always be attained.

Over the wire there come two more tubes not running the whole length but meeting and overlapping somewhat near the middle, so that at one point there are actually four concentric tubes besides the wire.

At the rear end a kind of cap called the breech-piece covers over the ends of all the tubes, itself having a central hole into which fits the breech-block, one of the triumphs of modern engineering, of which more in a moment.

While we have in mind the wire-wound form of construction it is interesting to note that something similar but in a crude form was practised sixty years or more ago. The guns of that era were some of them even of cast iron while the more refined consisted of a steel tube strengthened with coils of wrought iron. This iron was first rolled into flat bars, then it was made hot, and wound on spirally round an iron bar the same size as the tube. A little hammering converted this spiral into a tube which was then fitted round the steel tube. Thus, although very different there is still a distinct resemblance between this old method and the up-to-date wire-wound weapon.

The manufacture of guns, it may be remarked, owes more to one man than to any other, namely, Mons. Gustave Canet, a French engineer who, having fought in the Franco-German War, decided to devote his engineering talents to developing the artillery of his native land. He spent many years in England but later established works at Havre for the manufacture of guns upon improved methods, finally merging his interests into those of the great French armament firm of Schneider of Creusot. By French and English artillerists at all events the name of Canet is regarded with reverence.

But to get back to our naval gun. It will be clear that operations such as have been described, involving the handling of great tubes fifty feet or more in length, heating them as required, dipping them in oil while hot and so on, can only be carried out at works specially designed for the purpose.

The furnaces where the tubes are heated are well-like formations in the ground, deep enough to take the tube vertically. To lift them in and out there have to be tall travelling cranes capable of catching the tube by its upper end and lifting it right out of the furnace so that its lower end clears the ground. To accomplish this with a little to spare the cranes need to be seventy feet or so high.

Then there are deep pits full of oil so that a tube can be heated in a furnace, drawn out by a crane and quickly dropped into the adjacent oil bath. Likewise there have to be pits of a third kind wherein a cold tube can be set up while a hot one is dropped over it for the purpose of shrinking the latter on.

Then, of course, there have to be lathes of gigantic dimensions capable of taking a length of nearly sixty feet and of swinging an object weighing anything up to fifty tons. But of those machines we can only pause to make mention, for we must pass on to the breech-block, in some ways the most interesting part of the gun.

When it was first suggested to leave the back end of the gun open so that the powder and projectiles could be put in that way instead of through the muzzle, people at once foresaw how much would depend upon the arrangements for stopping up the hole while the gun was fired. For, of course, the force of the explosion is exerted equally in all directions, backward just as much as forward, so that unless very securely fixed the stopper closing the breech would be liable to become a projectile travelling in the wrong direction. To fix such a thing securely enough to avoid accidents would surely take up too much time and so largely neutralize any advantage arising from its use. These fears were, indeed, to some extent justified by accidents which actually occurred with the early examples of breech-loading guns, and for that reason our own authorities for a time looked askance at breech-loaders.

Now let us take a look at the breech-block of the 12-inch naval gun of to-day, which never blows out, not even when 350 lbs. of cordite go off just the other side of it. The explosion hurls an 850-pound shell at the rate of 8000 feet per second but it never stirs the breech-block. Yet it can be opened and closed so quickly, including the necessary fastening-up after closing, that shots can be fired from the gun at the rate of one every fifteen seconds.

The breech-block partakes of the nature of a plug and also of a door. It swings upon hinges like the latter but its shape more resembles the former. If we want to make such a thing very secure we usually make it in the form of a screw with many threads, but that entails turning it round many times and that takes time. Given plenty of time to screw the breech-block into its place and there would never have been any anxiety as to the possibility of its blowing out, but there is not time. The problem, therefore, was to get the strength of a screw combined with quickness of action.

This dilemma is avoided in the following simple manner. The breech-block is given a screw thread on its exterior surface, and the hole in the breech-piece is given a similar screw-thread on its inner surface, just as if the one were to be laboriously screwed into the other after the manner of an ordinary screw in machinery. Then four grooves are cut right across the threads on the block and similarly on the breech-piece, so that at four different places there is no thread left. In other words, instead of the thread running round and round continuously, each turn is divided up into four sections with sections of plain unthreaded metal in between. Thus in a certain position the block can be pushed into the hole without any threads engaging at all, for each strip of threaded block passes over an unthreaded strip in the hole and *vice versa*, in other words, the threads on the one part miss those on the other part. Yet an eighth of a turn serves to make all the threads engage and the thing is held almost as securely as if it were just an ordinary screw with threads its whole length.

The block is carried upon a hinged arm so that although it can be turned in this manner it can also be swung back freely when necessary.

Combined with the breech-block is a pneumatic contrivance which blows a powerful jet of air through the gun every time the breech is opened, thereby cleaning away the effects of the last explosion.

Each of these great guns is mounted upon a slide so that when it is fired it can slide back, thereby exhausting the effect of the recoil, yet can be returned instantly to its original position. Indeed, this return is brought about quite automatically by the agency of springs, compressed air and hydraulic power. Thus the gun fires, slides back, returns and is at once ready for the next shot.

It is trained, or pointed in a horizontal plane, by turning the turret in which it stands but the correct elevation is gained by the use of telescopic sights.

The principle of these sights is very simple. Imagine a graduated circle fixed to the side of the gun. Pivoted at the centre of the circle is a small telescope. The telescope can be turned round to any angle upon the circle and it can then be clamped at that particular angle.

The range having been given to the officer in command of the gun from the range-finding station on another part of the ship, the telescope is set to the correct angle. Then the gun is elevated or depressed until the ship being aimed at is precisely in the centre of the field of view of the telescope, in other words, until the telescope is pointing exactly at the ship. Then the gun is fired.

The effect, therefore, is this. The telescope always points (while the gun is being fired) at the object aimed at, but the gun is pointed upwards at a certain angle, which angle depends upon how the telescope is set upon the divided circle. Thus the setting of the telescope for a given range produces the correct upward tilt of the gun for that range.

The breech-block carries a trigger and hammer arrangement whereby the firing can be done and also an electrical arrangement so that an electric spark can be employed. Both these firing contrivances are so made that they cannot be operated until the breech-block has been inserted and *made secure*. Thus a premature explosion is guarded against.

CHAPTER 10

Shells and How They Are Made

Modern warfare seems to resolve itself very largely into a question of which side can procure the most shells. During the great war there was a time when the British and their allies were hard pressed because they had not sufficient shells. The enemy had in that matter stolen a march upon them and had during the winter, when military activity is at its minimum, rapidly produced large supplies of high-explosive shells.

Discovering their lack the British set about remedying it in true British fashion. It is quite characteristic of this strange people to let the enemy get ahead at the commencement, after which they pull themselves together and put on a spurt, so to speak, and after that the enemy had better prepare for the worst, for defeat is only a question of time. So, finding themselves short of shells, they set to and dotted the whole country in an incredibly short time with huge factories entirely devoted to making shells. Older factories also were adapted to the same purpose. Places intended and normally used for the manufacture of the most peaceable things—ploughs, gramophones and piano parts for example—were soon turning out shells or parts thereof by the thousand. Electric-light works, waterworks, cotton mills, technical schools, all sorts of places where, for doing their own repairs or for some similar reason, there happened to be a lathe or two, all these were organized and in a few weeks they too were working night and day *something to do with shells*.

Meanwhile other factories were springing up for the purpose of making explosives while others again were erected for producing the acids and other chemicals necessary for the explosive works; and yet another kind of works, the filling factories, came into being as if by magic and thousands of girls flocked from far and near to these places, there to fill the shells with the explosives.

BOMB THROWING

ONE OF THE MOST STRIKING THINGS ABOUT THE WAR WAS THE RE-INVENTION
OF THE BOMB THROWN BY HAND. THIS OFFICER HURLED BOMBS AT THE ENEMY FOR
TWENTY-FOUR HOURS CONTINUOUSLY.

Even the soldiers did not realize a few years ago how important the supply of shells was going to be. The rifle has fallen from its old place of importance while the gun and the shell have risen to the first place.

What, then, is a shell? It is what its name implies, a case covering something else, just as the shell of a fish covers its owner. It is a hollow cylinder of steel with certain things inside it. Its chief function is to hold these other things and to be shot out of a gun carrying them with it to their destination. You want to cause an explosion in an enemy's ship. You cannot get near enough to put the explosives there by hand, for he will not let you, so you put them into a steel shell and then hurl the whole thing at him out of a gun.

In the attempt to prevent your doing him any harm by thus throwing boxes of explosives at him, the enemy clothes the sides of his most valuable and important ships with thick steel plates, wherefore you have to make your shell strong and tough so that it shall not splinter against the armour but shall on the contrary bore its way through, finally exploding in the interior of the ship.

If it is not a ship that you are attacking but, say, an earthwork or an arrangement of trenches, then you do not need to penetrate steel armour and your shell can be thinner and of lighter construction. It still needs to be strong, however, for it has another function besides simply carrying the explosive. It must hold the force of the explosion in for a moment while it gathers force so that when the hour comes the pent-up energy may strike all round with the utmost violence. Even the most powerful explosives are comparatively feeble if they go off in the open. By holding them in check for a moment and then letting their force loose suddenly you get a much more forceful blow.

Shells which contain only an explosive are called common shells or high-explosive shells. Shrapnel shells constitute another type in which the force of the explosion is simply employed to release a number of round bullets, which strike mainly because of the velocity which they derive from the original motion of the shell. These are above all things man-killing shells, for their result is akin to a volley of bullets at close range.

We can thus sum up the chief types of shell as follows: the naval shell which has to be capable of penetrating armour: the high-explosive shell which must be able to break up earthworks and blow down the walls of trenches: and the shrapnel shell which scatters a shower of bullets and is most useful in attacks upon bodies of men rather than upon material structures.

Some shells have their propellant explosive combined with them just as the familiar rifle cartridge contains the propellant combined with the bullet. In the larger sizes, however, it is much more convenient to have the propellant in a separate cartridge, which can be handled separately and loaded into the gun separately.

As has been already explained, the propellant is a *powder* which gives a steady push rather than a destructive blow: moreover, it is practically smokeless, so as not to give away the position of the gun to the enemy. The high explosive, however, shatters and usually makes a dense smoke, so that the observers can see where it fell and report to the gunners whether or not they have got the range. Soldiers' letters have told us of the *black Marias* and *coal boxes* used by the Germans, those terms being simply soldiers' nick-names arising no doubt from the fact that certain particular shells are filled with *tri-nitro-toluene* which gives a black smoke. Clearly, smoke, which is most objectionable in the propellant, is a positive advantage in the bursting charge.

And now let us take a glimpse at the manufacture of one of these terrible missiles. An ingot of shell-steel is first cast as described in an earlier chapter. Since impurities are apt to rise, while the metal is liquid, the top of the ingot is always cut off and discarded. This waste material is used for many other purposes, in which a chance flaw would not be a serious matter, under the title of *shell-discard* steel.

The lower part is then heated and passed through a rolling mill, a machine very similar in principle to the domestic mangle, the rollers being of iron with suitable grooves cut in them. A few passages through this machine transforms the ingot into a thick round bar. This is then sawn into short pieces called billets, each of which is the right size to form a shell. Again heated, a powerful press drives a pointed bar through the softened steel, thereby converting the short billet into a rough tube. Another press then slightly closes in one end, making it resemble a bottle without a bottom and with the neck broken off.

The rough forging is then ready to be machined, an operation which is performed in a lathe. The outside is made perfectly round and smooth and of precisely the right size. The inside is also bored out to the correct diameter and finished off to an exceeding smoothness so as to avoid the possibility of any rough places irritating the explosive which in due time will be filled into it. For the same reason, the inside, when finished, is varnished in a certain way and with a certain varnish. The formation of this varnish is one of those little thought of but highly important services which alcohol renders to us, as mentioned elsewhere.

The smaller end (that which has already been partially squeezed in) is bored out and screwed for the reception of the nose-bush, while the other end is recessed for the reception of the plate which forms the bottom.

Most of these operations have to be very accurately carried out and, to ensure that that is so, gauges are continually employed to check the work. These gauges are based upon a very simple principle, known as the *limit* principle. This is both interesting and important, sufficiently so to merit a more detailed reference.

It must first be realized that no two things are alike and no measurement is perfectly correct. When we lightly speak of two things being alike we really mean that for the purpose contemplated they are nearly enough alike. Two things might be alike for one purpose and yet be so unlike as to be useless for another.

What the authorities do in the case of shells, therefore, and what is done nowadays in many branches of engineering, is to recognize this fact and at the same time overcome the difficulty by stating what difference is permissible. In other words, instead of saying that a thing must be a certain size, it is required to fall between two limits: it must not be more than one or less than the other.

For example, suppose a hole is required to be nominally an inch in diameter it may be specified that it shall not exceed an inch plus one-thousandth or fall short of an inch minus one-thousandth. In such a case a variation of a thousandth of an inch either way is permitted. The permitted variation may be more than that, or it may be less and be measured in ten-thousandths, it all depends upon circumstances. Clearly in every case it is desirable to permit as large a variation as is consistent with a good result. Now to make measures with the degree of accuracy just mentioned is not easy. One can just about see through a crack a thousandth of an inch wide if held up to a bright light. How then can dimensions such as these be dealt with easily and quickly in the rough conditions of a large workshop?

Let us again think of that one-inch hole and we shall see how simply and easily it is done. The gauge in such a case would be shaped somewhat like a dumb-bell, one end being the *go* end and the other the *not-go* end. The former is made to agree as nearly as possible with the lower limit, the other with the higher limit, and all the inspector has to do is to try first one end in the hole and then the other. One must *go* in and the other must *not-go*. So long as that happens he knows that the hole is correct within the prescribed limits. If, on the other hand, both go in, then he knows the hole is too large, or if neither goes in he knows it is too small. It may be urged by some acute reader that the gauges themselves cannot be correct, and that is quite true, but it is possible, by great care and laborious methods, to produce gauges which are correct to within far narrower limits than those mentioned. In the case of outside dimensions the gauges take the form of a thumb and finger capable of spanning the object to be measured, and in that case also two are used, one of which must *go* and the other *not-go*. By methods such as these the shells are measured and examined.

One of the most important features of a shell is its driving band. In the old days of round cannon balls it is said that the gunners used to wrap greasy rag round each so as to make it fit the cannon and to prevent the force of the explosion to some extent wasting itself by blowing past the ball. That is one of the functions of the driving band. It is made of copper which is comparatively soft, and it forms a fairly tight fit in the bore of the gun, so that while the shell is free enough to slide out of the gun it is tight enough to prevent the loss of any of the driving force of the explosive.

Its second purpose is to give the necessary spinning action to the shell. The old cannon ball suffered from the fact that it offered a considerable surface to the air in proportion to its weight. The idea arose, therefore, of making projectiles cylindrical and with a pointed nose, so that while the weight might be increased the resistance to the air might be even reduced. But it was clearly no use doing this unless the thing could be made to travel point foremost. Now for some rather mysterious reason, if you shoot a cylindrical object out of a gun, it will turn head over heels in the air, unless you give it a spinning motion. This motion, however, because of a gyroscopic effect, keeps the shell point first all the time.

It has another effect, too, known as *air-boring*. A spinning shell seems actually to bore its way through the air. Probably this is due to a centrifugal action, the spinning shell throwing the air outwards from itself and so to some extent sucking the air away out of its own path. Whether that be the true explanation or not, the fact remains that the spinning shell makes its way through the air better than a non-spinning one would do.

The gun, therefore, has formed in its bore a very slow screw-thread called *rifling*, from a French word meaning a screw. And it is the second function of the copper band to catch this rifling and by it be turned as the shell proceeds along the barrel. The soft copper conforms to the shape of the rifling and so itself becomes in a sense a screw engaging with the rifling.

This band is situated near the base of the shell, lying in a groove turned in the shell for its reception. To prevent the band turning round without turning the shell there is a wavy groove turned in the bottom of the larger groove, and the band, being put on hot, is squeezed into the latter by a powerful press.

The nose-bush is a little fitting of brass which screws into the smaller end of the shell and it has a hole in its centre into which another brass fitting, the nose itself, is screwed.

The base of the shell is closed with a little disc of steel plate. People sometimes wonder why the original forging is not made solid at the bottom so as to save the necessity for this disc, but the reason is that if that were done defects might very possibly arise in the steel in the centre which, since it is the very spot whereon the propellant acts, might let some of the heat or force of the propellant through, causing a premature explosion of the charge inside the gun itself instead of among the ranks of the enemy.

In the case of naval shells, the nose is not of brass but of a soft kind of steel. One might expect it to be of the very hardest steel, since it has to pierce the hard armour, but experience has shown that the soft nose is better than a hard one. The reason probably is that a hard nose splinters, whereas a soft one spreads out on striking the armour and then acts as a protection to the body of the shell behind it. In these shells, too, the fuse which explodes the charge is placed in the base. In the others it is in the nose, but clearly it could not be so placed in the armour-piercing shell.

It is interesting to mention that the propellant *powder* has combined in it some Vaseline or other greasy matter which acts as a lubricant between the gun and the shell when firing takes place.

Shrapnel is so different from the other types of shell that it merits a short paragraph or two to itself. Instead of being filled, as the others are, solely with explosive, the front part of it accommodates a considerable number of small round bullets, behind which comes a charge of gunpowder. The front half of the shell is separate from the back part, the two being connected by rivets of soft iron wire, so that a sudden shock can rend them apart. The shell is fired from the gun and comes flying along: suddenly, owing to the action of the fuse, the gunpowder explodes: the case then flies in two, the bullets are liberated and fall in a shower. In the South African War, where fortifications were few, these shells were very effective, but against fortifications, and particularly against trenches and barbed wire, big explosive shells are of much greater value.

CHAPTER 11
What Shells Are Made Of

The body of a shell is made of steel of a fairly strong variety. That is to say, it is stronger than that used for shipbuilding and for bridges and such work: but it is less so than some of the higher grades of steel, such as that used for making wire ropes. Owing to so much of this steel being rolled during the war, shell quality has come to be as well known to the general engineer as any of the many varieties which he has been accustomed to since his apprentice days. Many people wondered, at one time, why the cheaper and more easily worked cast iron could not be used for shells. There was a period when the steel works were quite unable to cope with the demands for steel, yet the iron foundries were crying out for work. This question then arose in many minds, Why not make cast iron shells? The answer is that cast iron is too weak: it would blow into fragments too soon.

Just think what a shell is and what it has to do. It is a metal case filled with explosive. It is thrown from a gun and is intended to blow itself to pieces on arrival at its destination. It is that self-destruction which carries destruction to all around as well. It is necessary, in order to obtain the best result, that an appreciable time should elapse between the ignition of the explosive and the bursting of the case. The force of the most sudden explosion is not really developed at once, but takes an appreciable time. After ignition, therefore, as the explosion gradually becomes complete, the pressure inside the shell is growing, and too weak a shell would go to pieces before the maximum pressure had been attained. Thus much of the energy of the explosion would simply be liberated into the air instead of being employed in hurling the fragments of shell with enormous force.

That is, of course, not a complete explanation of the whole action of a high-explosive shell, but it indicates generally the reason why a special quality of steel is required in order to get the best results.

Steel having been dealt with in another chapter, we will pass to the other metals which play important if not essential parts in the production of modern projectiles. So important are several of these that the lack of one or two of them would, under modern conditions, mean certain defeat for a nation.

Let us first of all take copper, of which is made the driving bands of the shells and which in combination with zinc forms brass of which noses and other important parts are made.

Its ore is found in many parts of the world, notably in the United States, Chile and Spain. The ores are of several kinds, the simpler ones to deal with being oxides and carbonates of copper, meaning compounds of copper with oxygen and with oxygen and carbon respectively.

It will be remembered that ores of iron are usually of the same nature, namely, oxides and carbonates, and consequently we find that the method of obtaining copper from these ores resembles the methods employed to obtain iron from its ores.

The ore is thrown into a large furnace, like the blast furnaces of the ironworks, and in the heat of the fire the bonds between copper and oxygen are loosened and the superior attractions of the carbon in the fuel entice the oxygen away, leaving the metal comparatively pure.

Unfortunately, however, copper is found most plentifully in combination with sulphur with which it forms what is termed sulphide. This copper sulphide is called by miners *copper pyrites*. Another trouble is that mixed with the copper pyrites there is usually more or less of iron pyrites, or sulphate of iron, so that to obtain the copper not only has the sulphur to be got rid of but also the iron. This complicates the operations very much, the ore having to be subjected to repeated roastings and meltings during which the sulphur passes off in the form of sulphur dioxide (a material from which sulphuric acid can be obtained), leaving oxygen in its place. Thus the copper sulphide becomes copper oxide, after which the oxygen is carried away by carbon, leaving the relatively pure metal. Moreover, at each operation various substances are thrown into the furnace called fluxes, which do not mingle with the metal but float on the top in the form of slag, and into the slag the iron passes, so that finally the copper is obtained alone.

Zinc is another important material for shell-making. Its ores used to be found in great plenty in Silesia, but the chief source of supply is now Australia. It is what is called *zinc blende*, and consists of *zinc sulphide*, or zinc and sulphur in combination. In all these names, it may be interesting to mention, at this point, the termination *ide* indicates a compound of two substances, so that we can safely conclude that the *ides* consist of the two elements named in their titles and no others. Thus *zinc sulphide* is zinc and sulphur and nothing else, *iron sulphide* is iron and sulphur, *copper oxide* is copper and oxygen, and so on.

The *blende* is first roasted in huge furnaces specially built for the purpose. To ensure its being thoroughly treated it has to be *rabbled* or turned over and over, since otherwise all of it might not be brought into contact with the necessary oxygen. At one time done by men with rakes, it is now generally accomplished by mechanical means.

A description of one such furnace will be of interest. It consists of a long rectangular building of brickwork bound together with steel framework. Inside it is divided up into low chambers, the roof of each forming the floor of the one above.

At intervals along its length mighty shafts of iron pass up from underneath right through all the floors, emerging finally above the topmost, while along underneath the furnace there runs a shaft the action of which turns the vertical shafts slowly round and round.

Attached to the vertical shafts are long strong arms of iron, one arm to each floor, and upon the arms are placed rabbles, as they are termed, pieces of iron shod sometimes with fireclay, resembling most of any familiar objects a small ploughshare.

As the arms slowly revolve, at the rate of once or twice per minute, the arms are carried round and round and the rabbles plough up and turn over and over the layer of ore lying upon the floor.

There are arms on the top of the furnace, too, sometimes, where the ore is first laid so that it may be dried by the heat escaping from the furnace beneath, an interesting example of economy effected by utilizing heat which would otherwise be wasted.

The whole of the furnace, from end to end and on every floor, is thus swept continually by the rotating arms with their dependent rabbles, and the latter are cunningly shaped so that they not only turn the ore over and over, but gradually pass it along the different floors or hearths. It is fed automatically by a mechanical feeder which pushes it on, a small quantity at a time, to the drying hearth on the top. Then the rabbles take charge of it and gradually pass it from the area swept by one shaft to that of the next until it has passed right along the top and has become thoroughly dried. Arrived there it falls through a hole on to the topmost hearth or floor, along which it travels by the same means but in the contrary direction until it again falls through a hole on to the top floor but one. And so it goes on until at last, fully roasted, it falls from the bottom floor of the furnace into trucks or other provision for carrying it away.

Some kinds of ore require to be heated by means of gas which is generated in a gas-producer near by. In others, however, the sulphur in the ore acts as the fuel, and so the furnace, having been once started, can be kept up for long periods without the expenditure of any coal at all. Very little attention is needed by furnaces such as these, so that with no fuel to pay for and very little labour, they are extremely economical.

Owing to the great heat, too, the arms would stand a very good chance of getting melted were they not kept cool by a continual stream of water flowing through the shafts and arms. This furnishes a continual supply of hot water which is sometimes used for other purposes in the works.

The process of roasting, whether carried on in furnaces such as these or not, results in the formation of oxide instead of sulphide; in other words, the sulphur is turned out and oxygen takes its place. The dislodged sulphur then joins up with some more oxygen and forms sulphur dioxide, which can be led away to the sulphuric acid plant and there, by union with water, turned into that extremely valuable substance, sulphuric acid.

We cannot, however, treat zinc oxide as we would iron oxide or copper oxide, for zinc is volatile, and so, instead of accumulating in the bottom of a blast furnace as the iron and copper do, would pass off up the chimney.

The oxide is therefore mixed with coal or some other form of carbon and placed in retorts made of fireclay. These retorts are fixed in rows one above the other like the retorts at a gasworks, and hot gases from a gas-producer down below pass around and among them. To the mouth of each retort is fitted a condenser, also made of fireclay.

Now what happens in the retorts is this: the heat loosens the bonds between the zinc and the oxide, the latter passing into union with some carbon from the coal. The zinc at the same time becomes vapour and passes into the condenser, the lower temperature of which turns it into a liquid which the workmen remove at intervals in ladles. On being poured into moulds and allowed to solidify this metal is called by the name of *spelter*, which bears to zinc the same relation that pig-iron does to the more highly developed forms of iron. *Spelter* is simply zinc in its crudest form.

Tin, although less important in war than copper and zinc, plays a not unimportant part. It has been found for centuries in Cornwall. The Romans used to trade with the natives of Britain for tin. Although considerable quantities of it is still obtained from there, the greatest tin-producing country of all at present is the Federated Malay States. Australia also furnishes ore, as does Bolivia and Nigeria.

In Cornwall the ore occurs as rock in veins or lodes filling up what must once have been fissures in granite rocks. That near the surface has long been taken, so that to-day the mines are very deep and costly to work. Some can only afford to operate when the market price of tin is above a certain limit. Much of the ore from the newer districts—the Malay States, for example—is in small fragments mixed with gravel in beds near the surface. Such is called alluvial or stream tin, since the deposits were undoubtedly put in their present position by streams or rivers. So long as they last these easily accessible alluvial deposits will always be cheaper to work than the deep mines. On the other hand, they may give out, and recent explorations underground seem to indicate that there is still much valuable ore not only of tin but of other metals too, to be obtained from the old mines of Cornwall.

The ore of tin, like so many other ores, is generally oxide. It is first roasted to expel sulphur and arsenic which are often present as impurities, and then it is melted in a reverberatory furnace such as that described for the manufacture of wrought iron. As usual, the oxygen combines with carbon, the impurities form slag which floats on the top, and the pure metal falls to the bottom of the furnace from whence it can be drawn off.

Mixed with or in the neighbourhood of tin ore there is sometimes found another mineral called wolfram, which plays an extremely important part in modern warfare, so much so that the British and other Governments engaged in the war were at times hard put to it to find enough. Its value resides in the fact that it contains tungsten, an element which has wonderful powers in hardening steel.

It consists of tungsten and oxygen, but is not an oxide since there is also iron in the partnership. This fact is very useful, however, since it enables the particles of wolfram to be picked out from the mass of other stuff among which they are found by a magnet.

There are some very wonderful machines called magnetic separators, made for this express purpose. In one, with which I am familiar, there is an endless band stretched horizontally upon two rollers. One of the rollers being driven round the belt travels along so that the mineral being fed on to it in a stream is carried along under several magnets. These magnets are very different from the ordinary magnet, inasmuch as they are revolving. We might almost describe them as small magnetized flywheels. As they spin round they pick up slightly the particles of ore which contain iron, but have no effect at all upon those which do not contain iron. They do not actually lift the particles up on to themselves: they just exercise a slight pull upon them, and by virtue of the fact that they are revolving, pull them off the band and throw them to one side. The wheels can be set closer or farther from the belt at will so as to make them act more or less strongly, and thus the most magnetic particles can be separated from those less magnetic, these latter being still kept separate from the wholly non-magnetic particles. Thus by simple and purely mechanical means are the precious bits of wolfram obtained from the other less valuable or worthless minerals with which they are mixed.

The same method is used with other minerals besides wolfram: it can be applied to all those which exhibit in some small degree the magnetic properties which we usually associate with iron.

This sorting out of one mineral from others continually crops up in connection with nearly all the metals except iron. Iron is practically the only one whose ore occurs in vast masses which need simply to be dug up and thrown into the furnace. The others, where they occur as rock in veins, have to be crushed to detach what is wanted from what is not wanted, and then the two have to be sorted in some way. Magnetic separation is but one of these ways. Another takes advantage of the fact that we seldom find two things together which have precisely the same specific gravity. Consequently, if we throw the mixture on to a shaking table the heavier particles will behave differently from the lighter ones and the two will separate. The same result can be obtained by throwing the mixture into a stream of water, the water acting differently upon the lighter and upon the heavier particles. Another way which may be mentioned is founded upon the fact that some things can be readily wetted with oil while others throw the oil off and refuse to be wetted by it. If a mixture of these two sorts of thing be stirred violently in a suitable oily liquid the former will be found eventually in the froth, while the latter will sink to the bottom. All these different methods are employed, as they are found necessary in preparing the ores of the various metals to which we have been referring.

Except in the case of alluvial ores which have been broken already by the action of ancient streams of water, nearly all ores (except iron) have to be crushed before the ores can be separated out. Some of this work is done by the very simplest contrivances, showing how in some cases invention has almost come to a stop through the machines having been reduced to their simplest form. A notable instance of this is the stamp mill, in which heavy timbers are lifted up by machinery and then allowed to slide down upon the ore, just like gigantic pestles.

More elaborate grinding machines are sometimes used, however, but it is impossible to mention them all here. The action of sorting out the fragments of ore from the miscellaneous assortment of crushed rocks is termed *concentrating*, and the sorted ores are called *concentrates*.

Another metal which has proved itself of immense importance in war is aluminium, and it fittingly comes at the close of the list since it is dealt with in a manner peculiar to itself. Practically all the others are obtained from their ores by means of heat and heat alone. Aluminium is obtained by electricity acting in the process called electrolysis.

It is surprising to learn that aluminium is one of the very commonest things on the face of the earth. Clay and many common rocks are very largely made of it. Clay, to be precise, is a silicate of alumina, a term which is interesting when it is explained. Silica is the name given to oxide of silicon. Sand is mostly silica. Alumina, too, is oxide of aluminium. Silicate of alumina is a combination of the two.

Any clay, therefore, could be used as an ore from which to obtain aluminium, but of course there are certain minerals specially suitable for the purpose, since in them the metal is plentiful and easily extracted.

In another chapter reference is made to the production of caustic soda from a solution of common salt by electrolysis. The same principle, precisely, is used to obtain the metal aluminium from its ore, which is generally an oxide.

Common salt, let me remind you, is sodium and chlorine combined. When you dissolve it in water it becomes ionized, which means that each molecule of salt splits up into two ions one of which is electrically positive and the other electrically negative. Then, when we introduce two electrodes into the solution and connect them to a battery or dynamo, all the positive ions go to one electrode and all the negative ions to the other.

We cannot dissolve aluminium ore in water, but we can in a bath of molten *cryolite*, and for some reason or other, whether because of the heat or not we cannot say, the ore becomes ionized, the aluminium atoms being one sort and the oxygen atoms the other sort. These ions then sort themselves out, the oxygen ions being taken into combination with the carbon rod which forms the positive electrode, while the metal ions collect upon the negative electrode. Since this latter is a slab of carbon which forms the bottom of the vessel in which the process is carried on, the result is that pure aluminium gradually accumulates in the bottom of the vessel and can be drawn off from time to time.

Aluminium is always produced in places where electric power can be obtained cheaply, such as near waterfalls.

CHAPTER 12

Measuring the Velocity of a Shell

In at least two of the preceding chapters of this book reference has been made to the speed at which a shell fired from a gun travels through the air. Such velocities as 3,000 feet per second have been mentioned in this connection, and some readers are sure to have wondered how such measurements could possibly be made. Possibly some sceptics have even supposed that they were not measured at all but simply estimated in some way or other. They are actually measured, however, and by very simple and ingenious means.

Needless to say, electricity plays a very important part in this wonderful achievement. In fact, without the aid of electricity it is difficult to see how it could be done at all.

People often ask how quickly electricity travels, as if when we sent a telegraph signal along a wire a little bullet, so to speak, of electricity were shot along the wire like the carriers of the pneumatic tubes in the big drapers' shops. That is quite a misconception, for in reality the circuit of wire is more like a pipe full of electricity, and when we set a current flowing what we do is to set the whole of that electricity moving at once. If we think of a circular tube full of water with a pump at one spot in the circuit, we see that as soon as the water begins to move anywhere it moves everywhere. Moreover, if it stops at one point it stops simultaneously at every other point. While practically this is the case it is theoretically not quite so, for the inertia of the water when it is suddenly started or stopped no doubt causes a slight distortion of the tube itself resulting in a very slight (quite imperceptible) retardation of the movement of the water. Electricity also has a property comparable to the inertia which we are familiar with in the objects around us, and there is also a property in every conductor which to a certain extent resembles the elasticity of the water-pipe, whereby it may for a moment be bulged out. In a short wire, however (up to a mile or so), particularly if the flow and return parts of the circuit be twisted together, this electrical inertia practically vanishes and consequently we may say that for all practical purposes the current starts or stops, as the case may be, at precisely the same moment in every part of the circuit.

That fact is of great value when, as in the case we are now discussing, we want to compare very exactly two events occurring very near together as to time but far apart as to place.

We need to compare the time when the shell leaves the gun with the time when it passes another point, say, one hundred yards away, and then again another point, say one hundred yards further on still. Supposing, then, a velocity of 3,000 feet per second, the time interval between the first point and the second and between the second and third will be somewhere about a tenth of a second. So we shall need a timepiece of some sort which will not only measure a tenth of a second, but will measure for us a very small *difference* between two periods, each of which is only about a tenth of a second and which will be very nearly alike. That represents a degree of accuracy exceeding even what the astronomers, those princes of measurers, are accustomed to.

This exceedingly delicate timepiece is found in a falling weight. So long as the thing is so heavy that the air resistance is negligible, we can calculate with the greatest nicety how long a weight has taken to fall through a given distance.

Near the muzzle of the gun there is set up a frame upon which are stretched a number of wires so close together that a shell cannot get past without breaking at least one of them. These wires are connected together so as to form one, and through them there flows a current of electricity the action of which, through an electro-magnet in the instrument house, holds up a long lead weight.

At some distance away, say one hundred yards, there is a similar frame also electrically connected to an electro-magnet in the same instrument house. This second magnet, when energized by current from the frame, holds back a sharp point which, under the action of a spring, tends to press forward and scratch the lead weight. The third frame is likewise connected to a third magnet controlling a point similar to the other.

To commence with, current flows through all three frames so that all three magnets are energized. The gun is then fired and immediately the shell breaks a wire in the first frame, cutting off the current from the first magnet and allowing the weight to fall. Meanwhile, the shell reaches the second frame, breaking a wire there, with the

BOMB-THROWERS AT WORK

MANY KINDS OF BOMBS ARE USED. ONE HAS A METAL HEAD AND A HANDLE ABOUT A
FOOT LONG, WITH A STREAMER TO ENSURE CORRECT FLIGHT; ANOTHER FORM RESEM-
BLES A BRUSH WHEN IT IS FLYING THROUGH THE AIR; AND A THIRD, KNOWN AS THE EGG,
IS OVAL IN FORM.

result that the second magnet loses its power, lets go the point which it has been holding back and permits it to make a light scratch upon the falling weight. This action is followed almost immediately by a similar action on the part of the third magnet, resulting in a second scratch on the lead weight.

The position of these two scratches on the weight and their distance apart gives a very accurate indication of the time taken by the shell to pass from the first screen to the second and from the second to the third. From those times it is possible to calculate the initial velocity of the shell and the speed at which it will move in any part of its course. Indeed, with those two times as data, it is possible to work out all that it is necessary to know about the behaviour of the shell.

This is rendered practicable by the fact that the moment the wire is cut the magnet lets go, no matter what the distance of the screen from the instrument may be. But for the instantaneous action of the current, allowance of some sort would have to be made for the fact that one screen is farther than another and the whole problem would be made much more complicated.

Even as it is, someone may urge that the magnets themselves possess inertia and will not let go quite instantaneously, but that can be overcome by making the magnets all alike so that the inertia will affect all equally. It is only necessary to have a switch which will break all the three circuits at the same moment (quite an easy thing to arrange) and then adjust all three magnets so that when this is operated they act simultaneously. After that they can be relied upon to do their duty quite accurately.

Thus by a method which in its details is quite simple is this seemingly impossible measurement taken.

CHAPTER 13

Some Adjuncts in the Engine Room

Before we deal with the subject of the engines employed in warfare, it may be interesting to mention two beautiful little inventions which have been made in connection with them.

Let us take first of all a contrivance which tells almost at a glance the amount of work which the engines of a ship are doing.

As everyone knows, there is in every ship (except those few which are propelled by paddles) a long steel shaft, called the tail-shaft, which runs from the engine situated somewhere near amidships to the propeller at the stern. Many ships, of course, have several propellers, and then there are several shafts. Now each of these shafts is a thick strong steel rod supported at intervals in bearings. If anyone were told that, in working, that shaft became more or less twisted, he would be tempted to think he was being made fun of. Yet such is literally the case. The thick strong massive bar becomes actually twisted by the turning action of the engine at one end and the resistance of the propeller at the other. And the amount of that twisting is a measure of the work which the engine is doing. The puzzle is how to measure it while the engine is running, for of course the twist comes out of it as soon as the engine stops.

A space on the shaft is selected, between two bearings, for the fixing of the apparatus. Near to each bearing there is fitted on to the shaft a metal disc with a small hole in it. On one of the bearings is fixed a lamp and on the other a telescope. When the engine is at rest and there is no twist in the shaft, all these four things—the lamp, the two holes, and the telescope—are in line. Consequently, on looking through the telescope the light is visible. But when the engine is at work and the shaft is more or less twisted one of the holes gets out of line and it becomes impossible to see the light through the telescope. A slight adjustment of the telescope, however, brings all four into line again, which adjustment can be easily made by a screw motion provided for the purpose. And the amount of adjustment that is found necessary forms a measure of the amount of the twisting which the shaft suffers and that again tells the number of horse-power which the engine is putting into its work.

But it is also necessary to know how fast the engine is working. There are many devices which will tell this, of which the speedometer on a motor-car is a familiar example. Most of those work on the centrifugal principle, the instrument actually measuring not the speed but the centrifugal force resulting from the speed, which amounts to the same thing. There is one instrument, however, which operates on quite a different principle, because of which it is specially interesting. It consists of a nice-looking wooden box with a glass front. Through the glass one sees a row of little white knobs. If this be placed somewhere near the engine while it is at work immediately one of the knobs commences to move rapidly up and down, so that it looks no longer like a knob but is elongated into a white band. There is no visible connection between the instrument and the engine, yet the number over that particular knob which becomes thus agitated indicates the speed of the engine.

Let us in imagination open the case and we shall find that the knobs are attached to the ends of a number of light steel springs set in a row. The springs are all precisely alike except for their length, in which respect no two are alike. Indeed, as you proceed from one side of the instrument to the other each succeeding one is a little longer than the previous one. Now a spring has a certain speed at which it naturally vibrates and other things being equal that speed depends upon its length. You can, of course, force any spring to vibrate at any speed if you care to take the trouble, but each one has its own natural speed at which it will vibrate under very slight provocation.

Every engine is, of course, made to run as smoothly as possible. All revolving or reciprocating parts are for this reason carefully balanced and in turbines the whole moving part, since it is round and symmetrical, naturally approaches a condition of perfect balance. Hence every engine ought to run perfectly smoothly. As a matter of fact, however, no engine ever does. There are certain limitations to man's skill and at the high speed of a fast-running engine, such as is to be found on a destroyer, for example, some little irregularity is sure to make itself felt by a slight vibration in the floor. It may be hardly perceptible to the senses, but to a spring whose natural frequency happens to be just that same speed or nearly so, it will be very apparent and in a few seconds that spring will be responding quite vigorously.

It is another example of the principle of resonance, which is employed so finely in making wireless telegraph apparatus selective. Every wireless apparatus is made to have a certain natural frequency of its own and it therefore picks up readily those signals which proceed from another station having the same frequency while ignoring those from others. In just the same way a reed or spring in this speed-indicator picks up and responds to impulses derived from the engine only when they are of a frequency corresponding with its own natural frequency. Hence, one spring out of the whole range responds to the vibrations of the engine while the others remain almost if not entirely unaffected.

In another form, the springs are actuated electrically. A magnet, or a series of magnets, is arranged so that as the engine turns the magnets pass successively near to a coil of wire, thereby inducing currents in that wire. They form, in fact, a small dynamo or generator, generating one impulse per revolution or two or three or whatever number may be most convenient. Then the current from this is led round the coil of a long electro-magnet placed just under the free ends of all the springs. The magnet therefore gives a series of pulls, at regular intervals, and the rapidity of those pulls will depend upon the speed of the engine, while the frequency of them will be registered by the movement of one or other of the springs.

This instrument can also be employed to determine the speed of aeroplane motors and, in fact, any kind of engine, especially those whose speed is very high.

CHAPTER 14

Engines of War

The phrase which I have used for the title of this chapter is often given a very wide meaning which includes all kinds and varieties of devices used in warfare. In this case I am giving it its narrower sense, taking it to indicate the steam-engines and oil-engines which are employed to drive our battleships, cruisers and destroyers, our submarines and our aircraft. They are inventions of the highest importance, which have played a large part in shaping modern warfare.

The type of engine almost invariably used on ships of war other than submarines is the steam turbine. Great Britain, for the most part, uses that particular kind associated with the name of the Hon. Sir C. A. Parsons, while the United States rather favour the Curtiss machine. Other nations have adopted either one of these or else something very similar.

All turbines are very simple in their principle, far more so that the older type of steam-engine, called, because the essential parts of it move to and fro, the *reciprocating* steam-engine.

In these latter machines there are a number of cylinders with closed ends and with very smooth interiors, in each of which slides a disc-like object called a piston. The steam enters a cylinder first at one end and then at the other, thus pushing the piston to and fro. The movement of the piston is communicated to the outside by means of a rod which passes through a hole in the cover at one end of the cylinder, the to and fro motion being converted into a round and round motion by a connecting-rod and crank just as the up and down motion of a cyclist's knees is converted into a round and round motion by the lower leg and the crank. The lower part of a cyclist's leg is, indeed, a very accurate illustration of what the connecting-rod of a steam-engine is.

As is evident to the hastiest observer, some arrangement has to be made whereby the steam shall be led first into one end and then into the other end of the cylinder: also that provision shall be made for letting the steam out again when it has done its work. Moreover, such arrangements must be automatic. Hence, every reciprocating engine has special valves for this purpose and such valves need rods and cranks (or something equivalent) to operate them. Further, to get the best results the steam must not simply be passed through one cylinder but through several in succession. Engines where the steam goes through only one cylinder are called *simple*, where it goes through two they are *compound*, where three *triple-expansion*, where four *quadruple-expansion*. Generally speaking, each cylinder has its own connecting-rod and crank, also its own set of rods, *etc.*, for working its valves. Hence, a high-class marine reciprocating engine is of necessity a complicated mass of cylinders, rods, cranks and other moving parts continually swinging round or to and fro at considerable speeds, all needing oiling and attention and all liable at times to give trouble.

And now compare that with the turbine, which has *two* parts, only one of which moves. That part, moreover, is tightly shut up inside the other one, being thereby protected from any chance of damage from outside and likewise rendered unable to inflict any damage upon those in attendance upon it.

At first sight it seems very strange that the turbine should be the newer of the two, for it is simply an improved form of the old time-honoured picturesque windmill which used to top every hill and grind the corn for every village and hamlet.

The old windmill had four sails against which the wind blew, driving the whole four round as everyone knows. The new turbine has a great many sails, only we now call them blades, and the steam blows them round. The old windmill had to have another smaller set of sails at the back for the purpose of keeping the main sails always in that position in which they would catch the full force of the breeze. In the turbine we need not do that, for we shut the windmill up in a kind of tunnel and cause the steam to blow in at one end and out at the other.

The difference between the various kinds of turbine lies simply in the manner in which the steam is guided in its passage through the machine.

After that general description we can take a more detailed view of the Parsons turbine. The casing or fixed part is a huge iron box suitably shaped for standing firmly and rigidly upon the floor of the engine-room. It is made in two halves, the upper of which can be easily lifted off when necessary. Often, indeed, this upper half is hinged to the lower, so that it can be opened like the lid of a box.

Inside, the casing is cylindrical, comparatively small at one end but increasing by steps till it is very much larger at the other end. At each end is a bearing or support in which the rotor or moving part is held and in which it can turn freely.

The rotor or part which rotates is a strong steel forging shaped somewhat to follow the lines of the inside of the casing. It does not entirely fill the casing but leaves a space all round and all the way along, which space is intended to accommodate the blades. The ends of the rotor are smaller than the body since they are intended to fit into the bearings, and one of the ends is prolonged so as to be available for coupling to the propeller-shaft of the ship.

At one end of the casing, the smaller one, is the steam inlet and the steam after emerging from it passes along till it finds its way out at a very large outlet formed at the bigger end. On its way it has to pass thousands of small blades so that the progress of each individual particle of steam is not a straight line but a continual zigzag. There are rings of blades round the rotor, tightly fixed to its surface. There are likewise rings of blades affixed to the inner surface of the casing, the rings upon the casing coming in the spaces between the rings on the rotor.

Let us imagine that we can see through the casing of a turbine at work and that looking down upon it from above we can trace the progress of a particle of steam. It rushes in from the inlet and at once makes straight for the outlet at the further end. Suddenly, however, it encounters one of the guide blades (those on the case) and by it is deflected to one side, we will suppose the left. That causes it to rush straight at one of the blades upon the rotor against which it strikes violently, giving that blade a distinct and definite push to the left. Rebounding, it then comes back towards the right but quickly is caught by another guide blade and by it hurled back upon a second rotor blade, giving it a leftward push just as it did to the first. Thus it goes zigzagging from one set of blades to the other until, tired out, so to speak, it finally flows away forceless and feeble through the outlet, having given up all its energy to the blades of the rotor against which it has struck in its course.

That, then, is the journey of one single particle. Multiply that by an unknown number of millions and you have a description of what takes place in the interior of a steam turbine. The blades are so proportioned, so arranged and so placed that it is very difficult indeed for a particle of steam to creep past without doing its share of work. Practically every one is made use of and while, of course, the action of a single particle of steam would have but a negligible effect, the vast number engaged cause the rotor to be powerfully blown round.

The reason why the casing and rotor are made larger and larger as one proceeds from the inlet towards the exhaust or outlet is that the steam must, if all its energy is to be extracted, expand as it goes and the enlargement provides room for this expansion.

One of the great advantages of the turbine is that the steam is always entering at the same end. In the cylinder of a reciprocating engine the steam enters alternately. It comes in hot but as it does its work and finally goes out it becomes very much cooler: the next lot of steam which enters, therefore, is chilled by the cool walls of the cylinder which have just been cooled by the departure of the previous lot of steam: so heat is wasted. Wasted heat means fuel lost, and as any given ship can only carry a limited quantity of fuel, wasted heat means less range and more frequent returns to the base to coal or to oil.

Also let me remark again upon the simplicity of the turbine as opposed to the other sort. The latter consists of a mass of moving and swaying rods and cranks, to work among which, as the engineers have to do, is a terrifying and nerve-racking experience. The turbine, on the other hand, has its only working part enclosed. It is difficult to tell, by looking at it, whether a turbine is at work or not, so silent and still is it, so self-contained. The reciprocating engine-room is noisy and full of turmoil: the turbine room is weirdly still by comparison.

On the whole, too, it makes better use of the steam which it uses, but it has one decided drawback. It will not reverse, which the other type of engine does readily.

This means that two turbines have to be coupled together, one with the blades so set that the steam drives it round correctly to produce motion ahead and the other set the opposite way so that it drives the vessel astern. The steam can be sent through either turbine at will and so motion can be obtained in either direction. Whichever turbine is in use the other revolves idly.

Unfortunately it is impossible to make a turbine to go slowly and yet be efficient. Consequently, slow steamers cannot use turbines, but for warships, which are nearly all fast boats, it has almost displaced the older type of engine.

The Curtiss turbine is different from the Parsons in that the steam encounters periodically, in its passage through, a partition perforated with funnel-shaped holes. Between the partitions it passes blades upon which it acts just as already described. The chief effect of this is to permit the machine being made of a rather more convenient shape and size. Other varieties of turbine are more or less combinations of the two ideas underlying these two.

When we look at a locomotive in motion we always see steam coming out of the funnel, but we never see that in the case of a steamer. That is because all the energy of the steam is taken and used in the latter case, while in the former much valuable energy goes off up the funnel, making a puffing noise instead of doing useful work.

On the steamship the steam is led not to the open air but to a vessel called a condenser the walls of which are kept cool by a continual circulation of cold water. The steam on entering the condenser at once collapses into water, leaving a vacuum. A pump called the *air-pump* removes the water (which was once steam) from the condenser and also any air which might get in, with the result that the engine is always discharging its steam into a vacuum. Thus to the pressure of the steam is added the suction of the vacuum.

In turbine ships the cooling water for the condensers is circulated by powerful centrifugal pumps driven by subsidiary engines.

The steam is obtained from boilers of that special variety known as *water-tube*.

The boilers with which most people are familiar are either Lancashire or Cornish, both sorts being large steel cylinders with two steel flues in the former and one in the latter running from back to front. The fire is made in the front part of the flue and the hot gases from it pass to the back and then along the sides and underneath through flues formed in the brickwork in which the boiler is set. Locomotive boilers, however, have no flues, but the hot gases from the fire in the fire-box pass through tubes which run from end to end through the cylindrical shell, each tube starting from the fire-box behind and terminating in the smoke-box in front. Thus we have tubes with fire inside and water outside: hence such boilers are called *fire-tube* boilers.

On many ships of the merchant type cylindrical boilers are used which combine the features, to some extent, of the Cornish and the fire-tube, since there is a flue running from front to back in which the fire is made and the hot gases return from back to front through a number of tubes which occupy the space above the fire. Arrived at the front the gases pass upwards to the chimney.

Water-tube boilers are different from all of these, since in them the water is inside the tubes while the fires play around the outside. This enables steam to be got up very quickly, a matter of much importance for a warship which may be called upon to undertake some operation at a moment's notice.

The boilers are fed with water from the condensers, so that the same water is used over and over again. When coal is burnt it is put on the fires by hand, for although mechanical stoking is a great success on land, there are special difficulties which prevent its use at sea. It is becoming more and more the fashion now to burn oil instead of coal in several types of ships and in those cases the oil is blown in the form of spray into the furnace. This has many advantages, some of which are exemplified on a small scale by the difference between using a coal fire and a gas stove. Like the latter, the oil spray can be quickly lit when needed and as quickly extinguished. It can be regulated and adjusted with equal facility. Oil can be taken on board too through a pipe, silently and quickly and without the terrible dirt and the exhausting labour involved in coaling a big ship. Oil, too, can be taken on board at sea, from a tank steamer, almost as easily as it can be taken in ashore, whereas the difficulty of coaling at sea despite many ingenious efforts has never been solved quite satisfactorily. Finally, oil can be stowed anywhere, for the stokers do not need to dig it out with a shovel. Therefore it can be carried in those spaces between the inner and outer bottoms which have to be there in order to give strength to the ship's hull but which would be quite useless for carrying coal. The advantages of oil fuel, therefore, are many and no doubt it will be used more and more as time goes on.

For Great Britain, oil fuel has the disadvantage that it has to be imported whereas the finest steam coal in the world is found in abundance in South Wales, but the difficulty may eventually be overcome by distilling from native coal an oil which will serve as well as that which is now imported.

So much for the turbine, the engine of the big ships: now for the Diesel oil-engine which drives the submarines. It belongs to that family of engines called *internal-combustion* since in them the fuel is burnt actually inside the cylinder and not under a separate contrivance such as a boiler. There have been oil-engines, so called, for many years, but they were really gas-engines since the oil was first heated till it turned into vapour and then that vapour was used as a gas. The Diesel engine, however, actually burns oil in its liquid state.

To understand how it works let me ask you to conjure up this little picture before your mind's eye. A hollow iron cylinder is fixed in a vertical position: its upper end is closed but its lower end is open: inside it is a piston, free to slide up and down: by means of a connecting-rod hinged to it and passing downwards through the open lower end the piston is connected to a crank and flywheel. At the upper end of the cylinder are certain openings which can be covered and uncovered in succession by the action of suitable valves.

Now let us assume that that engine is at work, the piston going rapidly up and down in the cylinder. As it goes down it draws in a quantity of air through a valve which opens to admit the air at just the right moment. The moment the piston reverses its movement and starts to go up again that valve closes and the air is entrapped. The piston continues to rise, however, with the result that the air becomes compressed in the upper part of the cylinder.

Now it is necessary to remind you at this point that compressing air or indeed any gas, raises its temperature. This air, therefore, which was drawn in at the temperature of the outer atmosphere, by the time the piston has reached the top of its stroke has attained a temperature well above the ignition point of the oil fuel.

The piston, having arrived at the top of its stroke, the upper part of the cylinder is filled with hot compressed air: the next moment the piston commences its descent, but at precisely that same moment a valve opens and there is projected into the cylinder a spray of oil. Instantly it bursts into flame, heating the air still more, so that as the piston descends the air, expanding with the heat, pushes strongly and steadily upon it. The amount of that push can be varied by varying the duration of the jet. The longer the jet is injected the more heat is generated and the more sustained is the push. On the other hand, if the jet is cut off very quickly the push is only a gentle one.

The power of the engine can thus be adjusted to suit varying circumstances by a slight variation in the valve which controls the jet.

The piston having thus been driven down to the limit of its stroke, it commences another upward movement, at which moment another valve opens and lets out the hot waste gases which have resulted from the burning of the oil. Thus the cylinder is cleaned out ready for a fresh supply of pure air to be drawn in on the next ensuing downstroke.

The engine thus works upon a series or cycle of operations which are repeated automatically over and over again. First comes a downstroke, drawing in air: then an upstroke, compressing it: then a second downstroke, during which the fuel burns and the power is generated: and, finally, a second upstroke during which the waste products of the burning are ejected. Power, it will be noticed, is only developed in one out of the four strokes: the other movements having, in single cylinder engines, to be performed by the momentum of the flywheel.

In most cases, however, the engine has several cylinders in which the cycles are arranged to follow in succession. Thus, if there are four cylinders, there is always power being developed by one of them.

The valves are operated automatically by the engine itself just as is the case with steam-engines. The engine also works a small pump which provides the very highly compressed air necessary to blow the oil jet into the cylinder.

Arrangements are often provided whereby the engine when working stores up a reserve of compressed air which can be used to start it. From the very nature of its working such an engine cannot develop power until it has accomplished at least four strokes or two revolutions, so that it cannot possibly start itself. If, however, compressed air be admitted to the cylinders to give it a vigorous push or two and so get it going, it can then take up its own work and go on indefinitely.

In some cases this is not necessary and that of an engine in a submarine is one of them. In that instance, the electric motor, which drives the boat when submerged, can be made to give the engine a start.

By altering the rotation in which the valves act the direction can be reversed. A very simple mechanism can be made to effect this change, so that reversing is quite easy.

Aircraft are mostly, if not entirely, driven by petrol engines, some of which are very little different from those of a motor-car or motor-cycle.

These motor-car engines are so well known that little need be said about them. It may be well to explain, however, that they, like the Diesel engines, work on a cycle of four strokes, as follows:—

First stroke (down) draws in a mixture of air and gas.

Second stroke (up) compresses the mixture. Just at the top of this stroke an electric spark fires the mixture, causing an explosion which drives the piston downwards, thus making the

Third stroke (down), during which the power is developed.

Fourth stroke (up) expels the waste products of the explosion.

Although all of them work on this same cycle, in which they resemble the engines of the motor-car, there are several much-used types of aero-engine in which the mechanical arrangement of the parts is quite different. Of these the best known is the famous Gnome engine which has a considerable number of cylinders arranged around a centre like the spokes of a wheel. The centre is in fact a case which covers the crank, and the cylinders are placed in relation to it just as the spokes are placed around the hub of a wheel.

There is only one crank and all the connecting-rods drive on to it.

Owing to their position around it they thus act in succession, giving a nice regular turning effort.

Further, these engines differ from all others in that the crank is a fixture while the rest of the engine goes round, exactly the opposite of what we are accustomed to. The engine, in fact, constitutes its own flywheel. Rushing thus through the air, the cylinders tend to keep themselves cool, doing away with the need for cooling water and radiators. Consequently engines of this type are the very lightest known in proportion to their horse-power. A fifty horse-power engine can be easily carried by one man.

It would be possible to go on much longer with this most interesting subject of engines, but having treated the three types which are most used in warfare, it is now time to pass on to something else.

CHAPTER 15

Destroyers

Except for the submarine the most prominent craft during the war has undoubtedly been the destroyer.

All warships are in one sense destroyers, since it is their prime duty to destroy other ships, so why should one particular kind of boat be given this name specially? Like many other of the terms which we use it is an abbreviation, a mere remnant of a fully descriptive title. Torpedo boat destroyer is what these ships are called in the *Navy List*.

Even that full title, however, only tells us what their original purpose was: it leaves us very much in the dark as to the many various functions which they perform.

The invention of the torpedo called for the construction of small boats whereby the new weapon could be used to best advantage, and so we got our torpedo boats. They in turn called forth another boat whose duty it was to run down and destroy them, and in that way we get our destroyers. From that bit of naval history we can almost see for ourselves what the characteristics of the destroyers must be. They have to be bigger than the torpedo boats, but as the latter were quite small the destroyers, though larger, are still comparatively small craft, latterly of about one thousand tons. Then they have to be very fast, in order to be able to chase the others and, finally, they need one or two guns, comparatively small so as not to overburden the ship and yet large enough to dispose of anything of their own size or smaller.

Unquestionably, their greatest feature is their speed. They are the fastest ships afloat, rivalling even a fairly fast train. Some of them can exceed forty miles an hour. They are very active and nimble, too, being able to turn in a comparatively small circle. For warships, too, they are cheap, so that a commander can afford to risk losing a destroyer when he would fear to risk another vessel. For all purposes except the actual hard-hitting they are the most useful weapon which the commander of the fleet possesses.

When the main fleet puts to sea a whole cloud of these smaller craft hover round looking for submarines or for the surface torpedo boats which might try to attack the large ships under cover of darkness, while keeping a sharp look-out, too, for mines or any other kind of floating danger, and thus they screen the more valuable ships.

Likewise do they convoy merchant ships sometimes, especially through waters believed to be infested with submarines. They also sally forth on little expeditions of their own, knowing that they can fight any craft equally speedy and show a clean pair of heels to any heavier ships, while by adroit use of their own torpedoes they may even bag a cruiser or two.

They are pre-eminently the enemy of the submarine, for the under-water boat is necessarily less active even when it is on the surface than they are, so that a submarine caught by a destroyer stands a very good chance of being rammed by it, which means that the destroyer deliberately rushes at it, using its own bow as a ram wherewith to knock a hole in it. Or if that be not practicable the destroyer, while dodging the torpedo of the submarine, may plant a single well-aimed shot into its opponent and the fight is over. A cleverly-handled destroyer appears to have little difficulty in avoiding the comparatively slow torpedo, but no ship ever built could avoid a properly aimed shell, two facts which are clearly indicated by the very few cases in which, during the war, a destroyer has succumbed to a submarine. The gun of the latter, if it has one, is no match for the guns of the destroyer.

Naval strategy and tactics, when one thinks about them carefully, reveal a very close resemblance to those of the football field. The destroyers are like the forwards, quick, light and nimble, valuable chiefly because of their ability to run swiftly and to dodge cleverly, while the heavy, stolid backs represent the battleships in their ability to withstand the heavy shocks of the game. Any imaginative boy will be able to carry this simile farther still and a comparison of the description of the battle of Jutland with his own knowledge of the game will reveal a surprising parallelism.

Thus the reader will to a very large extent be able to see for himself the manifold uses to which these wonderful little ships lend themselves, and he will see that above everything else it is their speed which counts, which fact gives us the key to their peculiar construction.

To commence with, they are made as light as possible. The material used is different from that of ordinary ships, being *high-tensile* steel, a steel into which a little more carbon than usual is introduced, resulting in about 50 per cent higher tensile strength but also involving, alas! rather more brittleness. When made of this material the whole framework of the vessel can be made of lighter beams and the covering can be of thinner plates than would be the case if the mild steel ordinarily employed for shipbuilding were used. The high-tensile steel is lighter for a given strength and therefore a ship built of it is lighter than it would otherwise have to be.

Besides the use of this particular material every resource in the way of ingenuity and skill on the part of the designers is bent towards saving weight. No unnecessary part is ever put in, but, on the other hand, necessaries are skinned down to the utmost limit consistent with safety in order to produce a light ship. How difficult this problem is is hardly realized until one thinks of the conditions which prevail when a ship floats in the water. The upward support of the water is exerted in a fairly regular way all along the ship while the weights inside which are pressing downward are concentrated in lumps. The engines, for example, represent a very heavy weight concentrated in one fairly confined spot. Thus the vessel has to have sufficient stiffness to resist the action of these opposing forces which are thus tending to break her in two. That, moreover, occurs in the stillest water; when the sea is rough still worse stresses are brought to bear upon the comparatively fragile hull, for a wave may lift each end, leaving the middle more or less unsupported, or one may lift the middle while the ends to a certain extent are left overhanging. All this, too, is in addition to the knocks and buffets caused by huge volumes of water being flung against the ship by cross seas in the height of a tempest. In the case of ordinary ships where speed is not of such great importance, the problem is simplified by the use of what is termed a high *factor of safety*, which means that the designers calculate these forces as nearly as they can and then make the structure *amply* strong enough. In other words, care is taken to keep well on the safe side. In a destroyer, however, there is no room for such a margin of safety. Risks have to be taken, and it is only the high degree of skill and experience possessed by our ship designers which enable these light ships to be made with, as experience shows, a very considerable degree of safety. They have to be continually choosing between strength on the one hand and lightness on the other and the way in which they combine the two is marvellous.

The weight thus saved is used for carrying engines, boilers and fuel. Relatively to its size, the destroyer is about as strong as an eggshell, but its engines are of extraordinary power.

The destroyers are generally organized and operate in little groups or flotillas of perhaps twenty or so with a small cruiser or a flotilla leader as a flagship, on which is the officer in command of them all. There is also usually a depot ship for each flotilla.

The flotilla leaders are what one might call super-destroyers, about double the size of the ordinary large destroyer, which is to say, about two thousand tons, and capable of very high speed.

The depot ships form a kind of floating headquarters for their respective flotillas. They are usually old cruisers which are specially fitted up for the purpose, and although they are of comparatively slow speed they can by wireless telegraphy keep in touch with the destroyers, which can return to them when occasion permits or demands. They carry workshops in which small repairs can be carried out, spare ammunition and stores of all kinds and spare men for the crews. In fact they can look after the smaller craft much as a mother looks after her children, and for that reason they are sometimes called mother ships.

As has been said, the destroyer was originally intended to destroy torpedo boats, but small torpedo boats have almost gone out of existence or rather the class have so grown in size as to have become merged in the destroyers, which, it must be remembered, are well armed with torpedoes which they have at times used with great effect. It is not surprising, therefore, to find that a still newer class of ship has arisen which has been described by one authority as *destroyer-destroyers*. Officially known as *light armoured cruisers*, not very much is known of their details. They are, however, about 3500 tons, with 10 guns, large enough that is to dispose of any destroyer which they might encounter.

Thus, to review the whole class of ships of which we have been speaking, we may say that there are the destroyers, all the more recent of which are about 1000 tons but diminishing as we go backward in time to about 350 or 400; the flotilla leaders about twice the size of the largest destroyers; and the destroyer-destroyers nearly twice as large as the flotilla leaders: all are characterised by high speed and by guns just large enough for the work for which they are intended. All are armed, too, with the deadly torpedo for attack upon larger ships than themselves.

They are essentially night-birds, much of their time being spent stealing about with all lights out, in pitch darkness, seeking for information or for a chance to put a torpedo into some chance victim. These night operations are very hazardous, but so skilful are the young officers who have charge of these boats that seldom do we hear of mishaps.

But although, as has been said, the torpedo boat has almost vanished, its under-water comrade has recently assumed a place in the first rank of importance, and perhaps to us the most valuable work of all done by the destroyer is that of hunting down and sinking these modern pirates.

CHAPTER 16

Battleships

Perhaps the greatest war invention of modern times was the British battleship *Dreadnought*. Of course, there have been battleships for centuries. In history we read of fleets consisting of so many ships of the line or in other words line-of-battle ships, meaning ships which were considered capable of taking their place in line of battle, as distinguished from frigates which correspond to the modern cruiser.

The line-of-battle ships were stout and strong with plenty of guns. They went into the thick of the fight, since they were capable of giving and receiving hard blows, while the lighter frigates hovered around seeking an opening to use their higher speed to cut off stragglers or to prey upon merchant ships.

Although so different in form and material that a sailor of the old days, could he revisit the earth, would not recognize them, the battleships of to-day are the real descendants of the line-of-battle ships of those times. They are stout and strong, with the heaviest guns, capable of giving and taking the hardest knocks, and it is they who form the backbone of the fleet. As we saw in the accounts of the battle of Jutland, the German Fleet tackled our cruisers and lighter vessels but discreetly withdrew when the battleships came up.

Looked at in another way, we may say that a battleship is a floating fortress. Its speed is not great, when compared with other ships, but it is constructed to carry enormous guns. It is also armoured with steel plates of great thickness and of special hardness placed upon the outside of the hull so as to cover its vital parts and protect them from the shells of the enemy. Its chief function, we may say, is to carry its guns: to enable it to do this with safety, it is armoured: and to enable it to get to grips with its enemies it has engines and boilers. Those are the three features of greatest importance in a battleship, its guns, its armour and its engines. All else is of minor importance.

It is strange to think how short a time the iron or steel ship has been with us. In the American Civil War, for instance, only about sixty years ago, the battleships were made of wood. It was during that war that Ericcson thought of the idea of putting iron plates to protect the sides of a ship from the hostile shots, and from that improvised armouring of a wooden ship has arisen the iron-clad or, more correctly, steel-clad monsters of to-day.

It is just about fifty years ago since the last iron-clad wooden battleship was launched for the British Navy. Her name was *Repulse*, and she took the water in 1868. With a tonnage of 6190 and a horsepower of 3350, she had a speed of 12 knots. Her armouring of iron was in parts 4½ inches and in other parts 6 inches thick, while she carried 20 guns of sizes which to-day would seem mere toys. If all her guns were discharged together she would throw a total weight of 2160 lbs. of projectiles.

Now, for comparison, let us take a modern battleship, the *Orion*, for example. The tonnage is 22,680, the horse-power 27,000.

She is more than twice the length of the older ship and is armoured with steel 12 inches thick. Her 10 large guns, each 13½ inches in diameter, if fired together (as I once heard them, like thunder, though 10 miles away) throw a weight of 12,500 lbs.

From this we see the wonderful growth in size, speed and in hitting power during the comparatively short period of fifty years. But there is a more striking comparison still.

The *Repulse*'s guns threw 2160 lbs. and the *Orion*'s throw 12,500. But that takes no account of the energy with which the weight is thrown. A tennis ball hit hard, might really contain more energy and do more damage to anything it hit than a cricket ball thrown gently, which illustrates the fact that in comparing the power of guns we need to consider something more than the mere weight of the projectiles. To arrive at a real comparison we take the weight of the projectiles in tons and multiply it by the speed at which they leave the guns in *feet per second*. And we call the answer so many *foot-tons*.

Now the energy of the *Repulse* thus reckoned comes to just under 30,000; that of the *Orion* to just under 690,000. The *Orion* can hit twenty-three times as hard as could its forerunner of only fifty years ago.

Since the *Repulse* all our battleships have been built of wrought iron or mild steel. Speaking generally, there was a steady development in size and horse-power and in speed until 1906, in which year there was launched the world-famous H.M.S. *Dreadnought*. Previously no battleship had been faster than 19 knots: she was designed for 21 knots. Her tonnage was 17,900, exceeding by more than 1000 tons anything that had gone before. But the great change was in the guns. Pre-Dreadnoughts had, or one ought to say *have* for there are still many in existence, four of the biggest guns, a number of medium-sized guns and a still larger number of smallish guns intended for the purpose of keeping off torpedo craft and such small fry.

At one stroke Lord Fisher, who was then the First Sea Lord of the British Admiralty, changed all this. He swept all the medium-sized guns away and gave this new ship *ten* of the largest guns then in use.

The advent of this ship startled the whole naval world, for it was seen at once by all those able to judge that there was a vessel which might be expected to sink with ease any other ship afloat. The on-slaught from those ten guns would be more than any other ship could stand. So other powers set to work to copy more or less exactly, while Great Britain quickly built more like her. So important was this new invention that very soon the strength of the naval powers began to be reckoned entirely on the number of Dreadnoughts they possessed, the older ships being left out of account as though they did not make any difference one way or the other.

But Great Britain was not content with the *Dreadnought*, for each succeeding ship or set of ships was improved until, only four years later, there was launched the *Orion* already referred to, nearly 5000 tons bigger, with 2500 more horse-power, and with 13½-inch guns instead of 12-inch. The *Orion* and her sisters are often spoken of as super-Dreadnoughts.

The Dreadnoughts as a class are often referred to as *all-big-gun* ships, since that is the feature which most distinguishes them from those which went before.

These large guns are mounted in turrets as they are called. We might describe these as turn-tables with a cover over something like a small gas-holder. There are usually two guns in each turret, although there are a few ships whose turrets have three in each.

The turrets seem to be standing on the deck of the ship and it is by turning them round that the guns are trained or pointed at their target.

The original *Dreadnought* had one turret in front and two behind, all on the centre-line of the ship, and two more, one each side, amid-ships. In late vessels all five turrets are on the centre-line. Thus the *Dreadnought* can fire six guns ahead, eight astern and eight to either side, while the newer ships can fire four ahead, four astern and all ten on either side.

There are other battleships with even more guns than these, such as the U.S.A. ship *Wyoming*, with twelve 12-inch guns, but the British Navy seems to prefer to stick to the original number of ten. The reason for this is that every such ship is a compromise between three alternatives.

The three great features have already been pointed out, namely, the guns, the armour and the propelling machinery. Either of these can be increased at the cost of one or both of the others, but all cannot be increased without sinking the ship, unless indeed, the ship be made larger and then other considerations crop up.

And that brings us to another class of ship often ranked among the battleships. These remarkable vessels are also termed cruisers and the fashion seems to have established itself of combining the two names and calling them battle-cruisers. They gave a fine account of themselves during the war.

The first three of these, of which the *Invincible* is usually taken as the type, made its appearance the year after the *Dreadnought*, and like the latter were the offspring of the fertile brain of Lord Fisher. The *Invincible* was about the same size as the *Dreadnought*, but had nearly twice the horse-power (41,000), which enabled it to attain an actual speed of nearly six knots more, namely, 28·6.

For guns it had eight of the same large weapons, and it was armoured with 7-inch steel armour-plates instead of 11-inch.

Thus we see illustrated what has just been said, less guns and thinner armour, to allow for more engine power and higher speed. Or, to put it the other way, we observe how higher speed was attained at the expense of the guns and the armour.

But just as the *Dreadnought* was followed by other still greater improvements in the same direction we get, in 1910, the famous ship *Lion*, a vessel not unknown to the Germans, a *super-Invincible*.

This ship has a tonnage of over 26,000 and 70,000 horse-power. It was designed to do 28 knots.

We saw the use of these ships in the Jutland battle, when, using their high speed, they attacked the German battleships and kept them engaged while the slower battleships came up. Though they suffered severe losses, which probably the more heavily armoured battleships would have escaped, they held the Germans so that it was only the failing light which saved them from utter destruction.

Another example was the way in which they hunted down Von Spee and his squadron off the Falklands, when they caught the Germans because of their higher speed and then sank them by means of their heavier guns with practically no loss to themselves.

We saw them again in the Heligoland battle, coming up to the assistance of the lighter vessels just in the nick of time and scattering the enemy like so much chaff.

A fact little known to most people and productive of much surprise is that these battleships and cruisers are not such very large vessels, when compared with those of the merchant service. The *Lion* is 660 feet long and 86 feet wide, the *Aquitania* is 930 feet long and 98 feet wide, and the *Olympic* is 882 feet long and 92 feet wide.

The mighty *Orion* makes a poorer showing still in point of size, since she is only 545 feet long and 88 feet wide—little over half the length of the *Aquitania*.

It is difficult to compare the tonnage of a warship with that of a merchant ship, since they are not measured in the same way. The former is the *displacement* or actual weight of water displaced: in other words the precise weight of the vessel in tons of 2240 lbs.

The tonnage of a merchant ship, however, has nothing to do with weight but is based upon capacity and is arrived at by a purely arbitrary rule, thus: all the enclosed space in the ship is measured in cubic feet and the total is divided by one hundred. That gives the gross tonnage. To arrive at the net tonnage the space occupied by the engines and all other space necessary for the working of the ship is excluded. Originally the tonnage of a merchant ship was the number of *tuns* of wine which it could carry.

Thus, you see, comparing the tonnage of a warship with that of a merchant ship is somewhat like comparing a pound with a bushel. Net registered tonnage is generally considerably less than the displacement tonnage of the same ship, so that a warship is usually less than a merchant ship of the same nominal number of tons.

And now let us turn to some of the internal arrangements of these wonderful ships, more particularly to the means for working the guns.

Each turret is placed over the top of what we might call a well, running right down deep into the inside of the ship. At the bottom of this well is the magazine, where the shells are stored and also the cartridges containing the explosive which drives the shell from the gun.

Underneath the turret, forming a kind of basement to it, is a chamber called the working chamber, and up to it the shells and cartridges pass by means of lifts. For safety's sake only a small quantity of explosives is kept here at any one time, but it is from here that the guns overhead are fed. Shells and cartridges alike pass up as required by means of hoists right to the guns. Indeed, the hoists are ingeniously contrived so that in whatever position a gun may be the hoist stops exactly opposite the breech, or opening at the back of the gun through which it is loaded. Then a mechanical rammer drives the shell or cartridge into its place in the gun.

The hoists are worked by hydraulic power or electricity, and in most cases by both, arrangements being made so that either can be used at will, thus serving as alternatives in case either should get out of order.

The turrets themselves are also turned by power. Indeed, so heavy are the weights involved that only by the use of carefully designed machinery is the operation of such great weapons made possible. A single shell of the 13·5-inch gun weighs 1250 lbs.

Around each turret there is placed a wall of thick armour plate as high as it is possible to make it without interfering with the movement of the guns. This is called the barbette armour and the space enclosed by it, in which the turret stands, is called a barbette, an old fortification term meaning a place behind a rampart.

The turret is covered over, as has already been remarked, by a steel hood, so that altogether the guns and their crews are about as well protected as it is possible to be.

That all this means a considerable burden upon the ship is shown by the fact that a pair of 12-inch guns with their turret and barbette armour will weigh something like 600 tons, and if there be five of them that means 3000 tons in all.

Down below in the magazine there are lifting appliances whereby the shells can be readily picked up and run to the hoist. Moreover, there is elaborate machinery for keeping them cool. Our allies the French had, years ago, several bad accidents through the explosives going off spontaneously in their ships, and this is quite likely to happen if the magazines become too hot. So refrigerating apparatus is installed similar to that employed in meat-carrying ships, which provides a constant flow of cool air into the magazines.

The ships also are subdivided to the greatest possible extent consistent with efficient working, so that in the event of a collision or a torpedo making a hole below water the ship may not sink. As far as possible the divisions or bulkheads are made to run right from top to bottom without any openings, but that obviously is a very inconvenient arrangement, so in many places there have to be doorways through them, leading from one part of the ship to another. In such cases these are closed by water-tight doors, which can be shut before the ship goes into action or into any dangerous region.

The engines of these vessels are now always turbines. This type of engine has many advantages over the older type, in which certain parts move to and fro, that motion being changed by cranks into a round and round action. For one thing, they are lighter for a given power, so that more power can be put into a ship without adding to the weight. That means higher speed. Then there is less to get out of order. Anyone who has been into a ship's engine room where to and fro or reciprocating engines are at work will realize this, for there is a maze of rods and cranks all moving together, and many parts which need to be oiled while in motion and which would get hot and tight if they were not carefully looked after. All this in an enclosed space with possibly an uncomfortable motion of the whole ship used to make the engineer's life at sea a very hazardous and unhappy one.

But the turbine is entirely enclosed. There is nothing to be seen moving at all. Indeed, there is only one moving part, and that is coupled directly to the propeller-shaft, so that nothing could possibly be simpler.

CHAPTER 17

How a Warship is Built

When it is decided to build a certain ship, the first thing to be done is to draw it on paper. The Admiralties of the world, and also the great shipbuilders, have each their own chief designer installed in a big, light, quiet office fitted with large strong, flat tables at which work a number of draughtsmen.

The naval authorities tell the chief in general terms what they want the ship to be capable of, and he determines its size and form. Then the draughtsmen work out his ideas on paper, themselves deciding upon the minor details, until they have produced exact representations of the ship which is to be. Some draughtsmen deal with the actual hull of the ship, while others design the various fittings and minor details, all working, of course, under the constant supervision of the chief.

In this connection one may perhaps allude to a matter which the general public often seems to misunderstand—the work and functions of a draughtsman. I have heard people say of a boy that he is good at drawing so they think of making a draughtsman of him. Now the point is that the actual drawing is perhaps the least important part of a draughtsman's work. He has to know *what to draw*. He is given just a rough idea of something and from that he has to produce a perfect design, bearing in mind that the thing to be made must well fulfil its purpose, must be easy and cheap to construct, must be strong enough yet not too heavy, must be made of the most suitable material and so on. He has to possess a good deal of the knowledge of the skilled workman, he has to be something of a scientist and a good mathematician in addition to his ability to make neat and accurate drawings. So, you see, these men whose minds conceive the details of our great ships are men of long training and experience, with far greater knowledge and skill than we sometimes give them credit for.

Anyway, there they stand, each at his own table, bending over his own drawing-board, each doing his own particular share towards producing the perfect ship.

But when all is said and done, there are limitations to the cleverness of the cleverest among us, so the next step, after the draughtsmen have done their best, is to test what they have done by experiment.

Years ago a certain Mr. William Froude interested himself in the question of the best shapes for ships, and he found that by making an exact model of a ship and then drawing that model through water it was possible to foretell just how that ship would behave. He built himself a tank for the purpose of these experiments at Torquay, where he lived, and by its aid he added a very important chapter to the science of shipbuilding.

Nowadays the Admiralty have a large and well-fitted tank at Portsmouth, the United States Navy have one at Washington, private shipbuilders have the use of a national tank at Bushey, near London, while several of the large firms have tanks of their own.

The national tank at Bushey, by the way, was given to the nation by Mr. Yarrow, a famous shipbuilder, in memory of Mr. Froude, it being called the *William Froude tank* in recognition of the great work done by him.

Now these tanks may be described as rather elongated swimming-baths. Such a structure is generally a little narrower than the average bath, but it is longer and much deeper.

At one end there are miniature docks in which the models float when not in use, while at the other there is a sloping beach upon which the waves caused by the models expend their energy harmlessly.

Along each side there runs a rail upon which are supported the ends of a travelling bridge. Driven by electric motors, this bridge can run to and fro from end to end of the tank, and its purpose is to drag the models through the water.

Carried upon the bridge is a platform which bears a number of instruments, chief among which is a self-recording dynamometer.

Now a dynamometer is an instrument for measuring the force of a *pull*, and when we call it self-recording we mean that it automatically takes a record of a series of pulls or of a varying pull. In this case there projects below the bridge a lever, to the end of which the model under test is attached. As the bridge rushes along it pulls the model through the water by means of this lever, and the force which is expended in doing so is recorded in the form of a wavy line upon a sheet of ruled paper.

If the model slips through the water very easily there is little pull upon the lever and the line drawn by the pen of the instrument remains low down upon the chart. If, however, much power is needed and the pull is a strong one the pen moves and the line rises towards the top of the paper. Any change, whether increase or decrease, is thus shown by the rise or fall of the ink line.

One model can be thus tried at various speeds and its behaviour noted under different conditions. Other matters can be investigated too, such as whether or not the bow rises in the water or falls when the boat is in motion, also how much such rise or fall may amount to.

The suitability of a certain shape of vessel, moreover, can to a certain extent be seen by observing the commotion which it makes in the water. Everyone has noticed the way in which a ship throws up a wave at its bows, and that bow-wave, as it is termed, represents so much energy being wasted. The power of the engines is absorbed to a certain extent in making that wave. It is impossible to make anything which when forced through the water will not make some wave, but certain forms cause less of it than others, and the designer of a ship seeks to find that form which will make the smallest bow-wave.

In like manner the eddies which a ship leaves in its wake are the result of wasted energy, and the ship must be so shaped that they too will be reduced to a minimum.

Shipbuilders find that there are three things which retard a ship's movement: skin friction, or friction between the water and the sides of the ship; wave making at the bow and eddy making at the stern. The first depends largely upon the smoothness of the ship's surface, the second and third depend upon its shape. If a model behaves badly in the tank the fault may be either too much wave making or too much eddy making, and which of these it is the dynamometer does not of course tell. In many cases the experienced eye of the tank officials furnishes the clue to the trouble, but in some cases a cinematograph is used to make a complete series of photographs of the model and the water around it as it rushes from end to end. These can then be studied in conjunction with the chart and the cause of the fault discovered.

The real aim, it is obvious, of all these tank experiments is to find out the lowest horse-power necessary to drive the ship, or the best form of ship to get the highest speed out of a given horse-power.

The cost of keeping up these large tanks and making the models and conducting the experiments is very great, for not only are the premises very large (I know one in which the water alone cost nearly a hundred pounds) but a highly skilled staff is necessary. The saving effected in the cost of ships and the superior efficiency of the ships makes it well worth while however.

There is still one other point about this matter which will possibly be puzzling the observant reader. What are the models made of and how are they made? They are made of paraffin wax, and a very important department of the experimental tank is that where the models are formed.

First of all a rough mould is fashioned by hand in modelling clay and into this is poured melted wax, the result being a very rough model of the ship. This is then placed in the model-making machine.

Those of my readers who are familiar with an engineer's shop will know what a planing machine is like, and from that they can form an idea of the general structure of this remarkable tool. There is, first of all, a travelling table which, as the machine works, travels to and fro. Spanning this table is a beam which carries on its under side two revolving cutters, so that as the table passes beneath them the cutters can operate upon anything placed upon the table.

Another part of the machine is a board upon which is placed the drawing showing the external shape of the proposed ship, and working over this board is a pointer connected by a system of rods and levers to the cutters just mentioned. The rough block of wax, then, having been placed upon the table and the to and fro motion set going, the attendant guides the pointer along the lines of the drawing, and as he does so the cutters so move as to carve away the soft wax into the precise shape of the model.

A little smoothing by hand is all that is necessary to complete the conversion of the rough piece of wax into a perfect model. It is then placed in the water and ballasted with little bags of shot until it floats at just the correct depth, and finally a light wooden frame is fitted to it for the purpose of making the connection to the lever by which it is pulled along.

Thus, after much thought and experiment, the designs for a new ship are completed. Tracings are then made of them on semi-transparent paper or cloth, which tracings are then used as *negatives*, from which a number of photographic prints are made, just as the amateur photographer makes prints from his negatives. At least that is how they used to be done, in a huge printing frame, but nowadays a machine is more often employed which passes the tracing or negative with a piece of photographic paper behind it slowly past an electric light, thus doing the work more quickly and more conveniently, for

THE TRIPOD MAST

HERE WE SEE ONE LEG OF THE TRIPOD MAST OF A WARSHIP. THESE MASTS HAVE GREATER STABILITY AND FREEDOM FROM VIBRATION THAN OTHERS. THEY ARE USED FOR OBSERVATION AND RANGE-FINDING, AND HAVE A FIGHTING-TOP ON WHICH GUNS OF SMALL CALIBRE ARE MOUNTED. HERE IS SHOWN A SAILOR CARRYING A WOUNDED COMRADE.

the drawings of ships are often very long and would either require an enormous frame or else would have to be made in pieces and joined together.

The prints are finally passed out to the works to be translated in terms of iron, steel and wood.

Perhaps the most important part of a shipyard is the mould loft, a large apartment on the floor of which the ship is drawn out full size. Then from these full-size drawings moulds or templates are made of wood or soft metal, showing the exact size and shape of the various parts. The moulds or templates go thence to the workshops, where the bars and plates of steel are cut to the right shape and perforated with holes, and some of the pieces are there joined together with rivets.

From the workshops the various pieces or parts go to the yard where the slip is on which the vessel is being built. This slip is by the water's edge, conveniently placed with a view to the fact that later on the great structure, weighing possibly thousands of tons, has got to slide down into the water.

Where the keel of the ship is to go a row of timber blocks is placed a few feet apart, and upon these blocks the plates of steel which form the lowest part of the ship are laid. Upon them are laid other parts, and upon them others, the joints being made by riveting. Thus the great ship grows from the keel upwards. As she gets bigger and bigger there comes the danger of her tipping over, and that is provided against by the use of props or shores along both sides.

By the time the hull is ready for launching it is often of great weight, all of which is borne upon the wooden blocks underneath the keel. Consequently, if the ground be not good, piles have to be driven in or concrete foundations laid to enable the huge mass of the ship to be supported. For this reason a large vessel cannot be built anywhere but only on a properly prepared *slip*, and it is the possession of a large number of such places which enables Great Britain to build so many ships at once.

Along each side of the slip there is usually a row of tall masts with a beam projecting out sideways near the top of each, forming cranes by which the heavier parts can be hoisted into position.

In other yards, again, there is a tall iron structure called a gantry along each side of the slip, while travelling cranes span across from one to the other over where the growing ship lies. These travelling cranes, worked by electricity, permit heavy weights to be handled with ease and safety. Other subsidiary cranes, meanwhile, carry the heavy hydraulic riveting machines by which riveting is done.

Much riveting is done by hand, men working together in squads of four. Of these one, often quite a boy, heats the rivets in a small furnace, after which he throws them one by one to man number two, who inserts each as he receives it in its proper hole and holds it there with a big heavy hammer or else a tool called a *dolly*. Number two is called the *holder-up*, since he holds the rivet up in its place while the remaining two hammer it over with alternate blows of their hammers.

In many cases, however, the two last described men give place to one, who is armed with a tool in shape much like a pistol and operated by compressed air obtained through a flexible tube. When he presses a trigger a little hammer inside the pistol gives a rapid series of blows to the rivet, completing the job more quickly than the two men can do with hand hammers.

A third way of doing this operation so important in the building of a ship is by the hydraulic machine suspended from the cranes. To the casual onlooker this has the notable feature of being silent, whereas riveting by hand and still more by a pistol hammer is terribly noisy. The reason for this is that the hydraulic riveter does not hammer at all, but, like a huge mechanical hand, it takes the rivet between finger and thumb and just squeezes it down.

One strange result of all this hammering in of rivets is that every ship by the time it leaves the slip has become a huge magnet, with somewhat disconcerting effects upon its own compasses, but of that more later on.

Thus the great ship grows, being made piece by piece in the workshops to the shapes indicated from the mould loft and put together and riveted on the slip, until finally in due time it is ready to take its first journey.

The launching of a big ship always strikes me as about the boldest and most daring thing which is ever done in the course of industry. For the huge structure, naturally top-heavy, weighing hundreds or thousands of tons, is just allowed to slide at its own sweet will. From the moment it starts until it is well in the water it is in charge of itself, so to speak, and if anything were to go wrong no power on earth could stop it once it had got a start.

That nothing ever does go wrong, or scarcely ever at all events, is due to the care with which all preparations are made before that critical moment when the ship is let loose and to the skill and experience of those in charge.

As the hull reaches that degree of completion when it can safely be put in the water, strong wooden structures termed launching ways are constructed one on each side of her. These really act like huge rails upon which in due course there will slide a gigantic toboggan. Tremendously solid and strong they have to be, as they have each to carry half the total weight of the ship.

Under each side of the ship and upon the launching ways there is built a timber framework capable of raising the ship bodily off the blocks upon which until now it has reposed. These two frames, being connected together by chains passing beneath the keel, constitute what is called the cradle, the *toboggan* which is to slide down the ways, bearing the ship upon it.

It is easy to see that being top-heavy something must be done to give the ship support before the shores on either side can be taken away, and it is equally clear that these latter must be removed before she can slide down to the water. Neither would it do to let the vessel slide upon her own plates, so we see that the cradle fulfils a twofold purpose, first enabling the ship to reach the water without ripping holes in her own plates, and secondly giving it the necessary side support to prevent it from toppling over on the way.

When all is ready, but a short time before the hour appointed for the launch, a curious operation is performed. Between the main part of the cradle and the part which actually slides upon the ways wedges are inserted, hundreds of them, and they are all driven in simultaneously. Their purpose is to make the cradle slightly higher and so to lift the ship off the blocks upon which it was built. If they were driven in one at a time each would only dig its way into the timber and nothing else would happen, but being driven all together a most powerful lifting action is produced which actually raises the mighty ship. So hundreds of men stand, each with his hammer ready to strike a wedge, while the foreman stands by with a gong. At a stroke on the gong the hundreds of hammers strike as one, and so the ship is raised off the blocks, which can then be removed, to facilitate which they too are built of wedge-shaped pieces which can easily be knocked apart. The shores, too, have ceased to serve any useful purpose and can be taken away until at last all shores and all blocks are gone and the vessel rests upon the cradle only. Meanwhile tons of grease have been put on the ways, and the ship, urged by its own weight, is straining to get down the greasy slope into the element for which all along it has been intended. At this stage the only thing which restrains it is a kind of trigger arrangement on either side which locks the cradle in its place. In some yards elaborate mechanical catches controlled by electricity are used for this, but in many the old device of *dog shores* is still used. These are simply two stout wood props which fit between a projection on the ways and one on the cradle, there being one dog shore on either side. Just over each dog shore there hangs a weight.

The person who performs the ceremony cuts the cord which holds the weights, the weights fall, the dog shores are knocked away, and the ship is free. Slowly at first, but gathering speed every moment, she moves majestically downwards into the water, being ultimately brought to rest by means of chains.

Whether done by the simple dodge of cutting a cord or by the more refined method of pressing an electric push, the launching is generally preceded by the breaking of a bottle of wine against the bows and the pronouncement of the vessel's name.

Once safely afloat, the vessel is towed away and berthed alongside a wharf whereon are cranes and other machines which lightly drop on board of her the massive turbines and boilers which in time will propel her, and the guns with which she will fight. All the multitudinous little finishing touches are here put into her until at last she sallies forth on her trial trips to show what she is capable of, after which follow trials of her guns, and then she takes her place in the fleet.

Thus, briefly sketched, we see the history of the warship from her inception in the minds of her designers till she is ready to meet the foe.

CHAPTER 18

The Torpedo

In parts of South America there lives a little fish, which, if you touch its nose, gives you a severe electric shock. The natives call it the *torpedo*. When an artificial fish came to be invented, capable of giving a very nasty shock to anyone touching its snout, that name was bestowed upon it too.

Even more than the submarine, the torpedo resembles a fish with its graceful outlines and its fins and tail, the chief difference being that the tail of the torpedo carries a couple of little rotating propellers. Looked at another way we may say that the torpedo is an automatic submarine. As a matter of fact, we all know it best as the weapon of the submarine.

It was originally invented by an Austrian who took it to a Mr. Whitehead, an Englishman who then had an engineering works at Fiume. This gentleman took up the idea and developed it into the Whitehead torpedo, which is to-day used by half the navies in the world, the rest using something very similar. It is curious to note that the German variety is called the *Schwartzkopf*, the meaning of which is *blackhead*.

The smooth, steel, fish-like body consists of two separate parts, which can be detached from each other. The front part called the *head* is made in two kinds, the war-head and the peace-head. The former contains a large quantity of explosive and the mechanism for firing it on coming into contact with any hard body. It is only used in actual warfare. The peace-head is precisely the same shape and weight as the other but is quite harmless, so that when it is fitted to the torpedo the latter can be handled with perfect safety, a valuable feature during the frequent exercises through which our sailors go in their efforts to attain perfection in the use and handling of these valuable weapons.

So much for the head. The body of the torpedo contains a beautiful little engine precisely similar to a steam-engine but on a small scale, which is driven by compressed air, a store of which is carried in a compartment provided for the purpose.

Then there is an automatic steering apparatus controlled by a gyroscope, the purpose of which is to keep the torpedo steered in precisely that direction in which it is started. If any outside force, such as current or tide, deflects it from its path the gyroscope, acting through a rudder at the tail, brings it back again.

Like the submarine, moreover, it has rudders which can steer it upwards or downwards and these again are controlled automatically so that having been set to travel at a certain depth the torpedo can be launched into the water with the practical certainty that it will descend to that depth and then maintain it.

This remarkable result is attained by the use of two devices acting in combination, namely, a hydrostatic valve and a pendulum. Either of these alone would set the thing going by leaps and bounds, at one time above the required depth and at another equally below it, and so on alternately. The hydrostatic valve consists of a flexible diaphragm, one side of which is in contact with the water outside, so that since the pressure increases with increasing depth, it is bent inwards more or less as the depth varies. This deflection is made to control the horizontal rudders. Suppose that things are adjusted for the rudders to steer the torpedo horizontally when at a depth of ten feet: if it descends to twelve feet the increased deflection of the diaphragm will so change the rudders that they will tend to steer slightly upwards: if, on the other hand, it rises to eight feet the contrary will happen, with the result that it will descend. As has been said already, this alone would result in a continually undulating course, so the pendulum is introduced to check the too decided changes in direction and so produce a practically straight course.

There is an interesting feature, too, about the propeller. It is *twin* but not, as in ships, two screws side by side. Instead, they are both set upon one shaft or rather upon two concentric shafts, like the two hands of a clock. The hour-hand of a clock is on one shaft, a solid one, which itself turns inside the shaft of the minute hand, which is hollow. The propellers of the torpedo are likewise, one on a tubular shaft and the other on a solid shaft inside it. These two shafts turn in opposite directions, but since the two propellers are made opposite *hands* they both equally push the torpedo along. The reason for this arrangement is that without it the action of a single propeller would tend to turn the torpedo over and over. Instead of the torpedo turning the propeller the propeller would to some extent turn the torpedo.

The range of the torpedo depends, clearly, upon the quantity of compressed air which it is able to carry and that is limited by certain practical considerations. One of these is the space required to store it, and a very ingenious method has been invented whereby the limited supply is eked out so that in effect its quantity is increased. As the air is used up the pressure in the air-chamber naturally falls and when that has gone on to a certain extent chemicals come into action which generate heat, whereby the remaining air is raised in temperature. This, of course, increases the volume of air and the result is just the same as if a greater quantity were carried to commence with.

The explosion is brought about by the pressing in of a pin which normally projects from the nose or point of the torpedo, and it would be very easy to knock this accidentally, causing a premature explosion, were not precautions taken to prevent it. These take the form of a little fan which is turned by the water as the torpedo proceeds through it. The firing-pin is locked by means of a screw so that it cannot be operated until it has been released by the withdrawal of the screw and that can only be done by the fan. Thus, while on the submarine or whatever ship carries it, the torpedo cannot be fired: it only becomes capable of explosion after it has passed through the water for a certain distance, far enough, that is, for the fan to have undone the screw. Thus the maximum of safety is combined with the maximum of sensitiveness when the object aimed at is struck.

There are other forms of torpedo which although little used are by no means lacking in interest. There is the Brennan, for example, at one time much favoured in the British Navy. Its propellers were operated from the shore, by the pulling of two very flexible steel wires. The effect was much as if the thing were driven by reins, as a horse is driven. On shore was a powerful engine with two large drums on which the wires could be wound and by which they could be drawn in at a very high speed. By pulling one more than the other the torpedo could be steered and it is said that such a torpedo could be made to follow a ship through complicated evolutions and fairly hunt it down, finally overtaking and striking it.

The purpose of such weapons was clearly to defend a port or roadstead against enemy craft which might try to rush in. It needed to be controlled by someone perched upon an eminence of some sort from which he could watch its course and guide it as might be necessary.

Compare this with the ease with which the Whitehead torpedo

is just slipped into the water and then left to itself. A submarine has in its bows either one or two tubes just large enough to hold the torpedo easily. At the front is a flap door which is kept closed while the torpedo is slipped into its place. Then the similar door at the rear of the tube is closed after which the front one can be opened. Water of course flows in and surrounds the torpedo when this takes place and a little push from some compressed air sends it floating out. As it emerges from the tube the engines are set going automatically and likewise the gyroscope which steers it, after which it continues to proceed in a straight line, soon seeking and maintaining the desired depth.

Other vessels besides submarines have submerged torpedo-tubes like these, but others again have tubes of a different kind. These are fixed on the deck and have the advantage that they can be pointed in any direction almost like a gun, whereas the others are either fixed rigidly in the vessel or are only slightly movable. In the case of these other tubes the torpedo is shot over the side of the ship, off which it leaps into the water somewhat like a man diving.

One other kind of steerable torpedo may be mentioned because of its ingenuity, although so far as is known it is not in actual use. It is called the *Armorl*, a compound of the names of its inventors, Messrs. Armstrong and Orling. It is controlled by wireless telegraphy in a very simple but effective manner. The rudder which steers it is connected to a small crank in such a way that as the crank revolves it turns the helm first to one side and then to the other. Suppose that, to commence with, the rudder is straight: a quarter of a revolution of the crank sets it to one side, say, the right: another quarter sets it straight again: a third quarter sets it to the left: and so on. The crank is turned by a wound-up spring, the effect of which is, however, normally held in check by a catch. When a wireless impulse comes along the catch is lifted for a moment, the crank slips round a quarter of a turn and the rudder is moved accordingly. Every impulse changes the position of the rudder and by sending suitable series of impulses it can be set as desired and changed at any moment.

A difficulty with all these guided torpedoes is that they must carry some indication whereby their place at any moment will be made visible to the man in control. A little mast and flag would do, for example, but it would be a fair mark for the enemy's guns and being shot away would leave the torpedo uncontrollable. The same objection seems to apply to the wireless antenna which this last type must carry with which to receive their guiding impulses, but that can be made light and almost invisible. It is when the thing is clearly visible that the danger arises, and, of course, to serve its purpose it must be visible. The way in which this difficulty was overcome by Messrs. Armstrong and Orling is a beautiful example of ingenuity. They cause a jet of water to be blown upwards by compressed air, something like the spouting of a whale, so familiar in books of natural history. That forms a mast which is clearly visible, yet the enemy may blaze away at it to their heart's content without damaging it in the least.

CHAPTER 19

What a Submarine is Like

The precise details of the submarines of our own navy or of any other for that matter are wrapped in mystery. Those who might tell do not know and those who know must not tell. True, there have been fully descriptive articles in many books and magazines, but it may be safely asserted that those descriptions are nothing more than what this chapter avowedly is, reflections by the authors on what such a craft must be like, more or less. It is just as well that this should be clearly understood, and the following description does not claim to be any more than that.

Just as an aeroplane follows the general design of a bird of the swallow type, which soars without flapping its wings, so the submarine necessarily follows much the lines of a fish. It has fins which help to guide it, it has rudders which compare with the fish's tail, and while it cannot use either fins or tail to push itself along as the fishes do, it has one or more propellers which serve that purpose admirably. It is rather remarkable that, while we often imitate nature very closely, there is one very important mechanical feature which almost invariably distinguishes man-made schemes from natural ones—that is, that man uses rotary motion for many purposes whereas nature practically never does. To be perfectly honest, the natural mechanisms are far too difficult for us to copy or I expect we should do so. For example, watch a goldfish and see how cleverly it uses its tail. Man could never hope to make anything so perfect as that tail. Absolutely under its owner's control, it serves a double purpose of propelling and steering in a manner which is equally beautiful and impossible to imitate.

For certain definite purposes, however, a rotary propeller is quite as good as anything which the fishes can show us. As a straightforward, simple, forward-pushing device it is equal to anything that a fish possesses. It has to be given that one duty, however, and no other, the steering being the task of a separate device, the rudder. There again, too, we see how nature does two things with one kind of mechanism while we have to use two, for the fish steers itself to right and to left with its tail in a vertical plane, but if it wants to steer upwards or downwards it twists its tail over somewhat towards a horizontal plane. The submarine, however, needs two distinct and separate rudders, one for right and left steering and one for up and down, the latter being generally a pair, one each side the vertical rudder for the sake of symmetry and balance.

So we find that a submarine has a body like that of a fish except that it is rather more rotund, perhaps, than the most portly fish usually seen. It has certain fixed fins projecting from its sides, which together with the rudders enable it to be guided. It has also certain long fins called bilge keels for the purpose of keeping it from rolling too much. Also, it has one or more propellers and the two kinds of rudder already referred to.

A fish, never wishing to get outside itself and walk about upon its own upper surface, needs no deck, in which the submarine differs from it, for the crew require somewhere where they can enjoy a breath of fresh air when opportunity offers. It is not a very commodious place, one could not exactly take a long walk upon it, nor even play deck-quoits, but on the back of the submarine there is an undoubted deck where the men can get out and upon which they can stand when she is on the surface.

A fish, moreover, takes little heed of things upon the surface: its interests lie almost entirely below. Hence it has no conning-tower or periscope, but without these the submarine would be useless. The former is a little oblong tower something like a chimney, which projects upward from the deck, while projecting to a higher level still is the tall hollow mast with prism and lenses at the top called the Periscope, through which the commander of the submarine, himself comparatively inconspicuous, can sweep the horizon for enemies or victims.

The problem of constructing a ship to travel under water is quite different from making one to travel on the surface in the ordinary way. When deep down the pressure of the water tending to crush the vessel is something enormous. Roughly speaking, it is a pound per square inch for every two feet in depth, so that if a submarine dives to a depth of fifty feet the water presses upon it with a force of about twenty-five pounds upon every square inch of its surface. On a square foot, that means over a ton. And there are many square feet of the surface in even a small submarine. Consequently, the whole shell of the ship has to be of very substantial construction. Moreover, there are curious strains which come upon the vessel when it dives to which surface ships are not subject. All these have to be reckoned as far as possible and allowed for.

The size of the modern submarine is not known with any certainty, but we may put it down roughly as two hundred feet long and at least a thousand tons displacement, which means that that is its actual weight, including everything and everybody on board, when it is just about to submerge.

Of course, a submarine, alone among boats, has two *tonnages*. When it is on the surface it is comparatively light. Indeed, *running light* is the technical term describing it when it is riding upon the surface of the water like an ordinary ship. Then, by increasing its weight, it can cause itself to sink until the little promenade or deck called the superstructure is just submerged and little can be seen above water except the conning-tower. That is termed the *awash* position, and it is clear that it is then displacing more water than when running light, and hence its displacement *tonnage* must be more.

When it is desired to sink, the vessel is set in motion in the awash position, from which it is gradually steered downwards by the diving rudders, until only the periscope, or it may be not even that, is left showing above. Then the maximum of water is being displaced. It is then actually displacing more than its own weight of water, for if left to itself it will rise rapidly and it is only the speed and the action of the rudders which keep it under. We see, then, that the action of a submarine in submerging itself is a real genuine dive. It sinks upon an even keel until it is awash, after which it goes under head-first, just as a swimmer does. It also rises bow first.

This tendency to rise when the combined action of movement and rudder ceases constitutes a very considerable safeguard, for should anything happen to the propelling machinery the vessel simply rises. At one time weights were attached to the under side of the hull which could be detached from the inside so that in the event of the vessel descending against the wish of her commander, she could be simply forced to the surface by the great excess of buoyancy resulting from shedding these safety weights. Of course, in the event of a serious perforation of the hull neither of these forms of surplus buoyancy would bring the boat up.

Let us now trace the operations of diving right through, supposing that our submarine is first running light. In that condition she is being driven by the oil engines which constitute her primary propelling power. The hatch or door at the top of the conning-tower is open, as also, it may be, is the one lower down, just at the foot of the tower. Men are standing upon the little platform formed by the tower, and one of them is steering by means of a wheel, keeping his eye, moreover, upon a compass also provided there, that being in fact, to the submarine when light, what the bridge is to the ordinary steamer. Other members of the crew may be upon the superstructure or deck just below, while others again are down inside, attending to their duties there.

Under these conditions the inside is by no means an unpleasant place. Plenty of fresh air comes down through the open hatches and through the ventilators, it being drawn down through the latter by means of a fan.

Preparations are then made for submerging. The hand-rail along the little deck is removed. The upper steering wheel and compass are covered up or shut away into the coverings provided for them, the wireless apparatus, if provided, is removed and the mast shut down. Hatches are securely closed and valves in the ventilating pipes are closed. In fact every opening is shut and made water-tight so that no risk shall be run of diving prematurely and taking in water accidentally.

The quarter-master transfers himself to the steering wheel inside, where he has another compass to guide him, not of the magnetic variety this time but a cunning application of the gyroscope. The commander, too, having descended before the last hatch was closed down, takes his stand at the eyepiece of the periscope, since that is now his only means of seeing what is going on above.

Another man takes his place at the wheel which controls the diving rudder, conveniently near to which is a pressure gauge so connected to the outer water that as the ship dives its depth is recorded upon its dial: that in effect is to him what the compass is to his comrade at the other wheel.

With every movement of men there needs to be adjustment made to keep the ship on an even keel. Otherwise she would go down by the bow or down by the stern according as the men's weight shifted towards either end. This is arranged for by two small tanks formed in the structure of the vessel, one at either end. Connected together by pipes and controlled by compressed air, water can be transferred from one to the other at will and so the balance be always kept. Quite simple manipulations of a valve serve to accomplish this delicate balancing performance. It is perhaps not of such importance at this stage, but in a moment, when the whole vessel will be under water, a very little movement indeed will suffice to upset the equilibrium.

143

Next water ballast is admitted into certain other spaces in the ship's structure, these spaces being called, because of the use to which they are put, ballast tanks. Gradually, as the incoming water increases the weight of the vessel, she sinks until she is awash. Then the diving rudders are set at the right angle (a pendulum serves to show the angle at which the boat points) and down she goes. As the pressure-gauge indicates the approach to the required depth the rudder is flattened out a little until just that position is found which keeps the boat under at the desired depth.

Of course, when all hatches and openings were closed the supply of fresh air was cut off and after that the crew had to depend upon the air contained in the submarine. Also, they had to stop the engine, for without air it cannot work: nor can it work without giving off fumes, which, if admitted to the ship, would soon suffocate the crew. Just before closing up, therefore, the engine is stopped and electric motors take up the task of driving the ship.

Now suppose that, while running submerged, the commander espies, through his periscope, an unsuspecting enemy. He tries forthwith to get as close as he can. Having noted the direction of the vessel and which way she is going and as far as possible her speed, he submerges more deeply, in all probability, lest the white streak which represents the wake caused by his periscope should reveal his presence. For possibly she is one of those terrible destroyers in fair fight with which he has but a poor chance. His only safety lying in complete invisibility, he therefore submerges entirely, trusting to his calculations to lead him in the desired direction. Thus he attempts and, if he have good luck, he succeeds in getting reasonably near to his foe.

Then he must try so to manoeuvre that his bow shall at the right moment be pointing towards the quarry, for his torpedo tubes are in the bow and they are fixed, or nearly so at all events, so that he can only fire them in a direction nearly, if not precisely, in the direction of the centre line of his ship.

Nay, he must do even more than that. It will not do to fire the torpedo directly at the ship, for a torpedo is comparatively slow. Suppose it is capable of forty miles an hour, and the other ship is a mile away: the torpedo will take ninety seconds to reach it. And in that time it may have travelled a mile or so itself. So the submarine man has to allow for that.

Occasionally, therefore, he comes up a little for a moment in the hope of getting a sight of the enemy while not revealing his own presence. Or perhaps he may decide to risk being seen and caught, trusting to the chance of getting his own blow in first. He needs to be a most resourceful man, with clear and keen judgment and supreme self-confidence, or he can never grapple with such a task.

Supposing, then, that he succeeds in getting undetected into a favourable position, as he thinks; at the critical moment the other ship may change its course, and the whole scheme goes awry. Perhaps he then tries to follow, but that is bad, for the end of a ship is not nearly so good a target as the side and the part hit is not so vulnerable. The first torpedo may, however, so disable the vessel as to give him chance to get into position for a second and better shot.

Anyway, when he thinks he has got his best chance he lets off a torpedo, immediately diving to be safe out of harm's way for a while. Then he rises to see the result of his work. If successful he would be sure to hear the sound, for water is an excellent sound-conductor and a submarine is like a gigantic telephone ear-piece.

It must be a nerve-racking job at the best of times, for the submarine is a very vulnerable craft. A member of the crew of a German submarine captured during the war is reported to have said that out of ten submarines attacked, nine were sunk. That may or may not be true, but it is certain that a very little damage, which would hardly affect an ordinary craft, is enough to sink a submarine. That is because, in order to be able to sink at will, the reserve of buoyancy has to be very low. An ordinary surface ship has at least as much of its bulk above water as below: hence it can take on board a weight of water almost equal to, if not exceeding its own weight before it sinks. At the best a submarine has not more than 30 per cent of excess and so it sinks if water amounting to only 30 per cent of its weight gets into it. In other words, the reserve in one case is at least 100 per cent: in the other at most 30 per cent.

During the war a submarine saw and tried to track down, somewhat after the manner described, a slow, steady-going collier which plies between London and the north carrying coal for a London gasworks. Having, as it thought, got into position for discharging its torpedo it rose for a final look when (it must have been to the amazement of the crew) the collier was seen making straight for them. What they really thought no one will ever know, for the collier had the best of the encounter, the submarine was crushed beneath her blunt bows and sank, no doubt, for ever. The mere fact that a slow, clumsy, heavily-laden collier could ever thus vanquish an up-to-date submarine is eloquent testimony to their vulnerability.

Many a submarine, too, has fallen to the shells of an armed fishing trawler simply because the shells of the latter were so much quicker in action than a torpedo, coupled with the fact that one well-placed shot, by preventing a submarine from diving, renders it almost helpless.

Some submarines, however, have a gun on the deck, so that when light they can fight like a destroyer or other lightly-armed vessel. The gun shuts down into a cavity when the vessel goes below.

The periscope, which forms such an important part of the submarine's equipment, is really very little more than a telescope. On the top there is a little mirror, or more probably a prism or three-cornered piece of glass which serves precisely the same purpose in that it reflects exactly as a mirror does. This is so placed that it throws the light from distant objects down the tube into the interior of the ship. In the tube are lenses very like those of an ordinary telescope and the light may be made to throw a picture upon a little table or screen or else can be viewed through another prism directly by the eye. In either case the periscope is just like an ordinary telescope set up vertically with a prism at the top so that it can *see* at right angles, and possibly another at the bottom so that the picture can be viewed at right angles to the direction of the tube. The latter is necessary only for the convenience of the observer, since otherwise he would have to be upon his back to look up the tube. The whole apparatus can be rotated mechanically and a scale forms a means of measuring the precise direction in which the prism or mirror is at any moment pointed. This is useful for measuring roughly the position of the prey, and it may even be used as a rough means of getting the range.

Another feature is the gyroscope compass, to which a passing reference has already been made. It is fairly well known that an object when spinning exhibits properties quite different from those which it possesses when still. A boy's top is a familiar illustration, for while spinning it will stand perfectly steady, supported only upon a tall peg with a sharp point, a pose which it will absolutely refuse to maintain when not spinning. Now fortunately for the present purpose it so happens that one of the peculiarities of the gyroscope or spinning-wheel is this: that if mounted in a certain way it persists in placing its axis in the same plane as that in which the axis of the earth lies. If you imagine for a moment a plane or flat surface of which the earth's axis forms a part you will see that wherever that plane cuts the surface of the earth will be a line in a north and south direction. Consequently, if any horizontal object has its axis in that same plane it, too, will always point north and south. A wheel, small but heavy, is therefore mounted with its axis supported horizontally upon a little metal raft floating in a trough of mercury and driven round at a very fast speed by a small electric motor fixed in it.

Whatever its position may be to start with, this revolving wheel will in a short time slew itself round upon the supporting mercury until its own axis is in the same plane as the axis of the earth: until, in fact, its axis points due north and south. Arrived in that position, it will remain there no matter how the ship upon which it stands may turn. Since it floats freely upon mercury the motion of the ship has little effect upon it, so little indeed, that it has no difficulty in following its own peculiar bent, even if the ship be describing circles.

The advantages of this are various: two of them may be stated. First, the apparatus points to the actual geographical north and not to the magnetic north, which is a slightly different direction and one, moreover, subject to frequent variation. Second, it is absolutely unaffected by the presence of iron or other magnets, a very fruitful source of error in the magnetic compass when used upon an iron ship close to steel guns and electrical machinery. Surrounded with iron as is the compass in the interior of a submarine, the magnetic needle practically refuses to work at all, so that, although employed on other ships, it is on the submarine that the gyro-compass finds its most important field of usefulness.

The pressure-gauge or manometer, which indicates the depth, is probably not different in any respect, except in its dial, which is marked in feet-depth instead of in pounds-pressure, from the pressure-gauge used on steam boilers. It has either a little cylinder with a piston in it which the water presses upwards more or less against the force of a spring, a diaphragm which is bent more or less, or a bent tube which tries to straighten itself out as the pressure inside it increases.

The older submarines derived their power from petrol engines similar to those which drive high-power motor-cars, but nowadays these have given place to engines of the type invented by the unfortunate Diesel who, after making one of the most brilliant and successful inventions of modern times, committed suicide, apparently in the height of his success.

These engines burn cheap heavy oil in place of the costly refined petrol: they are exceedingly reliable and well-behaved, and are free from many of the troubles which affect the petrol motor. They are referred to in more detail in another chapter.

In twin-screw boats there are two distinct engines, one for each propeller. Each engine, too, is coupled to a dynamo by which it can generate electric current, which is stored in large accumulator batteries until required and then withdrawn to drive the dynamos as motors while the boat is submerged, for if you feed a dynamo with current it becomes a motor.

A great deal of work is done, on the submarine, by compressed air, of which large stores are carried in strong steel cylinders. For example, the ballast is ejected from the ballast tanks, when the boat is required to rise, not by pumps but by the action of compressed air from a cylinder. The simple movement of a tap thus suffices to blow out the water in a very short time. The torpedoes, too, are given their initial push which sends them out of their tube into the water by compressed air. In other ways, too, compressed air is employed and to facilitate its use there are many tubes and valves whereby the cylinders and other apparatus are connected. Like all things human, these tubes and valves have their defects, which in this case means that they leak somewhat, but this defect is of value since the leaking air helps to keep pure and sweet the air inside the boat which, when submerged, the men have to breathe.

To what extent it is used I do not know, but it is a fact that certain chemicals, caustic soda for instance, have the power to absorb the objectionable carbonic acid which makes tightly-shut rooms seem close and uncomfortable, and if something of that sort be employed, it, together with the fresh air which thus leaks in by accident, is undoubtedly enough to enable men to live under water for many hours at a stretch.

On the other hand, several instances are on record in which strong healthy young officers have, after a course of service on a submarine, been found to be suffering seriously from chest and lung trouble, brought on, no doubt, by long spells of duty in this unhealthy atmosphere.

It used to be the custom to keep some white mice on board a submarine to give warning of the impurities in the air. Being very susceptible to the smell of petrol vapour, which used to be a source of considerable danger, and also to carbonic acid, these little creatures squeaked with anxiety some time before the conditions became really dangerous, thus giving timely warning. There is an instrument, however, which will give an indication of this sort and probably it has been brought in to reinforce the mice if not actually to supplant them. This interesting little instrument, which the gasworks people use for detecting leakage, consists of a metal drum with a porous diaphragm. Normally the pressure of the atmosphere upon the diaphragm is equalled and balanced by the pressure of the air inside the drum, but if there be gas in the air this balance is upset, the diaphragm is bulged in or out and a finger is thereby moved, which movement forms a measure of the amount of gas present.

In conclusion, we may fittingly take a glance at what happens when a submarine founders. Only a few years ago this occurred with lamentable frequency, though now it is quite rare except under the actual stress of warfare. Several interesting schemes were therefore invented to give the men at least a sporting chance of getting to safety. One was to make the conning-tower detachable and water-tight, so that the men could get into it, fasten themselves in and float up to the surface. The practical difficulties in the way prevented this being a success. For example, if sufficiently detachable in an emergency it was difficult to make it sufficiently water-tight in ordinary use.

Another and better device provided the men with small helmets and jackets, like the dress of a diver very much simplified. One of these for each man was stored in an accessible place in the boat and partitions were devised inside the hull itself in order that whatever happened there should be air entrapped somewhere wherein the men could live for a time and put on their helmets in safety. Then, thus provided, they could crawl out through the hatchway and float up to the surface. Arrived there they could inflate their jackets by blowing into them, open the window of the helmet and float upon the surface in comparative safety until rescued.

This apparatus was largely installed in British submarines and a tank was built at Portsmouth where the men could actually practise with it under water.

A third device may also be mentioned. This takes the form of a buoy fitted into a recess in the boat's upper surface. Sufficient line is coiled up inside it and when the occasion arises it can be released from inside. This does not in itself save the crew but it may go a long way towards ensuring their safety by letting those above know just where the sunken craft is and guiding them in their efforts to raise it.

The torpedo, the weapon without which the submarine would be practically useless, is dealt with in another chapter. Enough has been said here to give a good general idea of these interesting craft, their fittings, their uses and the sort of life which befalls those who man them.

CHAPTER 20

The Story of Wireless Telegraphy

For ages people were puzzled as to the nature of light. Pythagoras, that old Greek who invented what we now call the forty-seventh proposition of Euclid, thought that the bright body shot off streams of tiny particles which literally hit the observer in the eye. Sir Isaac Newton thought the same, but for once *the greatest scientist of all time* was wrong.

For when the Danish astronomer, Romer, discovered that light travelled at the rate of somewhere about 186,000 miles per second it dawned upon people that it was scarcely believable that particles of any kind could by any means be made to move so fast. So they set about searching for a new explanation, and they found it in the idea that light was conveyed from the bright body to the observer's eye by means of waves, and as there cannot be waves of nothing they had to imagine a something to exist in all the vacant spaces of the universe capable of forming the waves of light. This something was called the *luminiferous* ether or light-bearing ether. We can neither see, feel, taste nor hear it. Our senses tell us nothing about it. Indeed, if it does really exist it must be so very different from anything that we do know by our senses that one is often tempted to doubt its existence. Still, it explains so many things which are otherwise unexplainable and enables us so correctly to reason from one phenomenon to another that our reason forces us to accept it as a fact, at all events until something better comes along.

This wave theory in regard to light was finally set at rest by the curious discovery about a century ago by Dr. Thomas Young of London that if two lots of light were brought together in a certain way they produced darkness.

Now if a ray of light were a stream of particles, two such rays would inevitably and always, if added together, produce a doubly brilliant light, and under no conceivable circumstances could they do anything else. But two lots of waves can, and do, under the proper conditions, neutralize each other so as to produce rest.

This mutual action upon each other of two sets of waves can be very simply exhibited by two violin strings tuned to *nearly but not quite the same note*. If you have a violin handy, try it and you will find that when either string is plucked separately it gives a steady continuous sound, but if both be plucked at the same time they give a throbbing sound. That is because, periodically, as one string is coming up the other is going down, so that they neutralize each other, while at other times, owing to the fact that one is vibrating faster than its fellow, both are rising and falling together. When neutralizing each other there is a momentary silence, while in between the silences come the times when both are acting together and therefore producing a specially loud sound. And so as the vibrations of the faster keep gaining upon those of the slower string one hears a continual crescendo and then diminuendo repeated over and over again. So two sets of sound waves sometimes produce silence.

And in like manner two sets of light waves can be made so to *interfere* (that is the technical term) that together they produce darkness.

So for a hundred years or more people have, generally speaking, accepted the idea that light consists of waves in a medium called The Ether. Heat also is brought to us from the sun and from any distant hot body by similar means, the difference between light waves and heat waves being simply in their wave length or the distance apart. The different colours of light, too, are to be accounted for by different wave lengths.

You have of course seen how a magnet can act upon a piece of iron at a distance. You may, too, have tried the experiment of jerking a magnet past a piece of wire, thereby generating an electric current in the wire. Both those things need, for explanation, that we assume the existence of a something invisible and undetectable by our senses between the magnet and the iron and between the magnet and the wire, by which the action of one is conveyed to the other. So people imagined another Ether capable of acting like a link between the magnet and the iron and between the magnet and the wire.

Now just about half a century ago a celebrated professor of Cambridge University brought all these facts about light, heat, magnetism and electricity together and by skilful reasoning showed that but one Ether sufficed to explain all these things. He showed how magnetic and electric forces acting together could produce waves like those of light and heat. And finally he demonstrated by figures that waves so formed would necessarily travel at the very speed at which light and heat are known to move.

This is known as the electro-magnetic theory of light. And not content with showing the nature of things already known, Professor Clerk-Maxwell added a prophecy that there were other waves in existence of longer wave length, which no one then knew how to make or to detect if made.

Following up this prophecy many investigators sought these waves, and the first to find them was Professor Hertz of Karlsruhe in Germany. Fortunately for his position in the minds of English people he died before the War, so that his name is not sullied by the stupidities of which German professors in more recent days have been guilty. On the contrary, his writings show him to have been a kindly, modest, genial soul, and particularly gratifying is his generous assertion in one of his books that had he not himself discovered these waves he is certain Sir Oliver Lodge would have done so. He seemed quite anxious to share the credit of his discovery with his English colleague as he called him.

Let us see then how these *Hertzian* waves are produced. In the year 1748 a Dutch experimenter named Cuneus thought he would try to electrify water. He got a glass flask and filled it with water into which he let drop one end of a chain connected to an old-fashioned frictional electrical machine. Thus he stood with the flask in his hand while a friend worked the machine. After a short time the friend stopped and Cuneus took hold of the chain to lift it out, when to his astonishment he received a shock which knocked him over, broke his flask and sent him to bed to recover.

Unwittingly Cuneus had invented what became known thereafter as a Leyden jar, Leyden being the town in which he lived. It consisted, you will notice, of two conductors, the water and his hand, with an insulator, the glass, in between.

To understand or rather to give ourselves a useful working explanation of how such an apparatus comes to be charged we must first imagine that everything contains a certain normal amount of electricity which we can by certain means add to or take away from at will. When we add some to anything we say we have given it a positive charge: when we subtract some we say that we have imparted a negative charge. Clearly, if we add some to one thing we must first obtain it from something else, and if we take some away from one thing we must do something with what we have taken, and so we add it to something else. Therefore whenever we charge anything positively we must charge something else negatively and vice versa.

Now the ease with which we can thus charge two bodies seems to depend upon their nearness to each other, so that the easiest things to charge are two plates of metal separated by the thinnest possible insulator. Modern Leyden jars are usually formed of a thin glass jar with a lining inside and out of tinfoil.

The Leyden jar is, however, only one form of the piece of electrical apparatus known as an electrical condenser, and many other forms exist. For example, a flat sheet of glass with foil above and below, or several such piled one on top of another. An eminent electrician whom I know has recently made some of two tin patty pans put bottom to bottom, nearly but not quite touching, the whole being enclosed in a solid block of paraffin wax. And I might describe many other forms, but whatever they may be every one is essentially two conductors with an insulator between.

Now when a condenser has been charged its charges remain for a considerable time unless they be given a chance to escape. Suppose you have a charged condenser and that you take a wire and with it touch simultaneously both the conductors, the surplus on one plate will rush through the wire and make good the deficiency upon the other; it will thus in an instant become discharged.

Now several scientific men had suggested, before Hertz's time, that when that occurred something else happened too. They thought that the charge did not simply rush from one plate to the other instantly, but that it oscillated to and fro for a period; that the surplus rushing round overshot the mark, so to speak, and not only made up the deficiency but caused a surplus on the opposite plate, after which this new surplus rushed back again through the wire, doing the same thing, though to a less and less degree, several times over before a condition of perfect rest was reached. To use a simple analogy, it was thought that the surplus swung to and fro like the swinging of a pendulum. We know that a pendulum swings because of its inertia, and electricity possesses a property very like inertia which, it was thought, would cause it to behave in the same way.

The Ether waves travel at the rate of 186,000 miles per second, so that if, as was thought, a sudden current of electricity gives rise to a wave, currents which succeed each other at the rate of one per second would produce waves 186,000 miles apart. A hundred currents per second would give a wave length of 1860 miles. A thousand per second would give 186 miles. But a thousand succeeding currents per second are difficult to produce, and 186 miles is so very much greater than the tiny fraction of an inch, which is the length of the light and heat waves, that Hertz had to find some way of making currents succeed each other faster even than a thousand times per second.

So he thought of these oscillating currents which were supposed to occur when a condenser was discharged, and he rigged up a condenser with an induction coil and a spark gap in a way which he thought would do what he wanted.

There is not room here to explain the Induction Coil, indeed it is so well known that it will be quite sufficient to state that it is an apparatus which takes steady current from a battery and gives back instead a lot of little spurts or splashes of current at a rate of, say, fifty or one hundred splashes per second, according as we adjust the little vibrating spring which forms a part of the coil. We can so connect this to a condenser that each splash will charge it up; and we can combine with it a spark-gap, that is to say, a gap between two knobs, so that every time it is charged it immediately discharges again through this gap. Thus we may have, say, one hundred splashes per second, and each splash is followed by several oscillations across the air-gap, the oscillations taking place at the rate of perhaps a million per second. Each series of oscillations is called a *train*.

Now a million per second gives a wave-length somewhere about what Hertz wanted, so he arranged his apparatus as just described.

For a condenser he used two metal plates a little distance apart, the air between forming the insulating material. He set up his apparatus in a large room, and having started the coil he moved about with a nearly complete hoop of wire, the ends of which nearly touched. Working in darkness he found after a while that sometimes he could see little sparks, very small but just visible across the gap between the ends of the bent wire. Those sparks only occurred when the coil was in action, and so he knew that the one was the result of the other's work. By careful painstaking experiment he found that the sparks were unquestionably caused by waves, and that the waves moved with the same speed as light, also that they could be reflected and refracted just on precisely the same principles as those which control light. Moreover, he measured the wave-length.

At first sight it seems incredible that anyone could measure the distance apart of waves which travel at such a speed as 186,000 miles per second, but fortunately, by a special application of *interference*, it is possible to make the waves stand still and tamely submit to measurement. An example of this can be seen by simply tapping a glass of water, when

LISTENING FOR THE ENEMY
SPECIAL SENSITIVE CYLINDERS ARE SUNK INTO THE GROUND TO WHICH THE
USUAL TELEPHONIC APPARATUS IS FIXED. THIS ENABLES THE SAPPERS TO
DETECT ANY UNDERGROUND OPERATIONS BY THE ENEMY.

the ripples being reflected off the sides interfere with each other and become stationary. Stationary waves are half the wave-length of the original waves, and by using this method Hertz was able to make a measurement which at first sight seems beyond the bounds of possibility.

Thus Hertz discovered how to make the waves which Clerk-Maxwell had predicted and also how to detect them when made.

It was not long before the idea arose of using these waves for signalling to a distance. Many experiments were made but with no very striking success until 1896 when Marconi first came to England.

Hertz had noticed that the farther apart he placed the plates of his condenser the farther could he get his tell-tale spark, so Marconi saw that the plates of his condenser, too, must be far apart. He also found that the earth could be used as one of the plates, that in fact there was a great advantage in so using it. So, one plate having to be the earth itself and the other removed as far as possible from it, the tall masts of the wireless antenna came into being.

When Marconi came to England he was taken under the kindly wing of Sir William Preece, the veteran engineer of the Post Office, and the facilities which Sir William was able to give no doubt helped largely in his subsequent rapid progress. After a few experiments in London he got to work across the Channel, sending messages from the North Foreland Lighthouse to Wimereux on the coast of France, including congratulatory messages between the French authorities and good Queen Victoria.

A little later he was signalling from Niton in the Isle of Wight to the mainland and to the far west at the Lizard. The first wireless telegram which was actually paid for was sent by Lord Kelvin, the father of cable telegraphy, from Niton to the mainland, whence it was transmitted by land wires to Sir George Stokes. This incident, so interesting because of its marking a stage in the history of this great invention, also because of the persons concerned, occurred in 1898.

But Marconi was quickly increasing the range of his apparatus far beyond anything already mentioned. He journeyed in the Italian warship *Carlo Alberto* as far north as Cronstadt and as far east as Italy, keeping in communication with England all the time. Then he crossed the Atlantic, again keeping up communication with England the greater part of the journey.

Raising his wires to a great height by means of kites he was soon able to signal from Nova Scotia to the great station just previously built at Poldhu in Cornwall, and then wireless telegraphy from land to land across the great ocean became an accomplished fact.

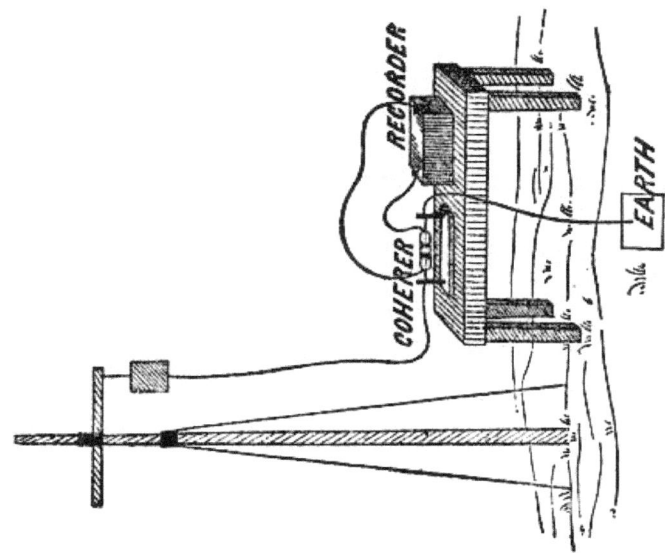

TRANSMITTER.

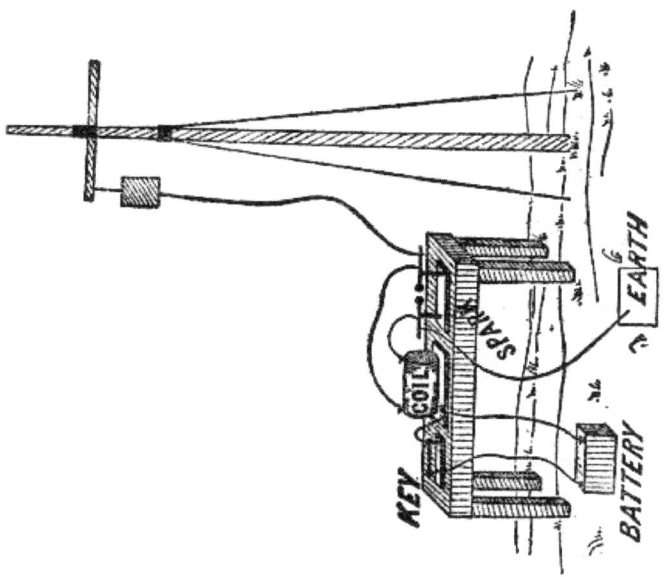

RECEIVER.

DIAGRAM SHOWING THE PRINCIPLE BY WHICH THE AERIALS ARE CONNECTED TO THE APPARATUS.

We all know how things have progressed since then. A telegram by Marconi is as commonplace to-day as a telegram by cable. The British Government is now engaged upon a series of stations dotted about the globe in such a way that every part of the widely separated British Empire shall be in constant touch with every other part by wireless telegraphy. In other words, the range of the system has now become such that nothing further is needed.

The British Admiralty has a few wires slung to posts on the top of the offices in London, and those few wires enable touch to be maintained with ships. As almost every intelligent newspaper reader in Great Britain knows, the Germans were in the habit, during the war, of sending news to the United States by wireless telegraphy, which news was always picked up by the Admiralty installation and circulated to the British newspapers, often to the amusement of their British readers.

The famous *Emden*, too, which had such a run of success until it encountered the Australian cruiser *Sydney*, met its end entirely through the intervention of wireless telegraphy.

These incidents give us a good idea of the usefulness of wireless in naval warfare. In military work it is used chiefly in connection with air-craft, but of that more will be said in another chapter.

Chapter 21

Wireless Telegraphy in War

The history of this wonderful invention has been described in the preceding chapter. Now we will see how it is applied in warfare.

Let us take first its uses in connection with the Navy. The aerial wires or antenna are stretched to the top of the highest mast of the vessel. Where there are two masts they often span between the two. Ships which have masts for no other reason are supplied with them for this special purpose. In the case of submarines, the whole thing, mast and wires included, is temporary and can be taken down or put up quickly and easily at will.

The stations ashore are equipped much after the same manner as are the ships, except that sometimes they are a little more elaborate, as they may well be since they do not suffer from the same limitations. For example, the well-known antenna over the Admiralty buildings in London consists of three masts placed at the three corners of a triangle with wires stretched between all three.

However these wires may be arranged and supported they are very carefully insulated from their supports, for when sending they have to be charged with current at a high voltage and need good insulation to prevent its escape, while, in receiving, the currents induced in them are so very faint that good insulation is required in order that there may not be the slightest avoidable loss.

The function of these wires, it will be understood, is to form one plate of a condenser, the earth being the other plate and the air in between the *dielectric* or insulator.

In the case of ships the *earth* is represented by the hull of the vessel. It makes a particularly good *earth* since it is in perfect contact with a vast mass of salt water, and that again is in contact with a vast area of the earth's surface. Salt water is a surprisingly good conductor of electricity.

In land stations *earth* consists of a metal plate well buried in damp ground. The whole question of conduction of electricity through the earth is very perplexing. There seems to be resistance offered to the current at the point where it enters the ground, but after that none at all. Consequently the resistance between two earth plates a few yards apart and between similar ones a thousand miles apart is about the same. Though the earth is made up mainly of what, in small quantities, are very bad conductors indeed, taking the earth as a whole it is an exceedingly good conductor. That makes it all the more important that where the current enters should be made as good a conductor as possible, and the construction and location of the earth plates is therefore very carefully considered so as to get the best results.

Wires, of course, connect the antenna to the earth, thereby forming what is called an *oscillatory circuit*. The ordinary electric circuit is a complete path of wire or other good conductor around which the current can flow in a continuous stream. An *oscillatory circuit* is one which is incomplete, but the ends of which are so formed that they constitute the two *plates* of a condenser. In that way, according to theory, the circuit is completed between the two ends by a strain or distortion in the *Ether* between them. A continuous current will not flow in such a circuit, but an alternating, intermittent or oscillating current will flow in it in many respects as if there were no gap at all but a complete ring of wire.

At some convenient point in this oscillatory circuit are inserted the wireless instruments, one set for sending and the other set for receiving, either being brought into circuit at will by the simple movement of a switch.

In small installations the central feature of the sending apparatus is an Induction Coil operated by a suitable battery or by current from a dynamo. Connected with it is a suitable spark gap consisting of two or three metal balls well insulated and so arranged that the distance between them can be delicately adjusted. This is generally done by a screw arrangement with insulating handles, so that the operator can safely adjust them while the current is on.

The current from the battery or dynamo to the coil is controlled by a key similar to those used in ordinary telegraphy, the action being such that on depressing the key the current flows and the coil pours forth a torrent of sparks between the knobs of the spark-gap, but on letting the key up again the sparks cease. Since the sparks send out etherial waves which in turn affect the distant receiving apparatus it follows that a signal is sent whenever the key is depressed. Moreover, if the key be held down a short time a short signal is sent, but if it be kept depressed for a little longer a long signal is sent, by which means intelligible messages can be transmitted over vast distances.

Certain specified wave lengths are always used in wireless telegraphy. That is to say, the waves are sent out at a certain rate so that they follow each other at a certain distance apart. In other words, it is necessary to be able to adjust the rate at which the currents will oscillate between the antenna and earth. Every oscillatory circuit possesses two properties which are characteristic of it. These two properties are known as Capacity and Inductance. It is not necessary to explain here what these terms mean precisely. It is quite sufficient just to name them and to state that the rate at which oscillations take place in such a circuit depends upon the combined effect of these two properties. Consequently, if we can arrange things so that capacity or inductance or both can be added to a circuit at will and in any quantity within limits, we can within those limits obtain any rate of oscillation which we desire and consequently send out the message-bearing waves at any interval we like; in other words, we can adjust the wave-length at will.

Fortunately, it is very easy to add these properties to an oscillatory circuit in a very simple manner. A certain little instrument called a *tuner* is connected up in the circuit and by the simple movement of a few handles the desired result can be obtained quickly even by an operator with but a moderate experience. He has certain graduated scales to guide him, and he is only called upon to work according to a prearranged rule in order to obtain any of the regulation wave-lengths.

As a matter of fact, the instruments are not directly inserted in the antenna circuit, the circuit that is which is formed by the aerial wires, the earth and the inter-connecting wires. Instead, the two sides of the spark-gap are connected together so as to form a separate circuit of their own, the local circuit as we might call it, and then the two circuits, the antenna circuit and the local circuit, are connected together by *induction*.

A coil of wire is formed in each, and these two coils are wound together so that currents in one winding induce similar currents in the other winding, and by that means the oscillations set up by the coil in the local circuit are transformed into similar oscillations in the antenna circuit. This transformation involves certain losses, but it is found in practice to be by far the most effective arrangement. Both the circuits have to be tuned to the desired wave length, but that is done quite easily by the operation of the handles in the tuner already referred to.

It is to this coupling together of tuned circuits that Marconi's most famous patent relates. It is registered in the British Patent Office under the number 7777, and hence is known as the *four sevens* patent. It has been the subject of much litigation, which proves its exceptional importance, and it is to the fact that the Marconi Company have been able to sustain their rights under it that they owe their commanding position to-day in the realm of wireless telegraphy.

The receiving apparatus also consists of a separate local circuit which can be coupled when desired to the antenna circuit through a transformer. The same simple tuning arrangement is made to affect this circuit also, so that the *multiple tuner*, as the instrument is called, controls all the circuits both for sending and for receiving. The oscillations caused in the antenna circuit by the action upon it of the *etherial* waves flowing from the distant transmitting station pass through one winding of the transformer and thereby induce similar oscillations in the local receiving circuit which are made perceptible by the receiving instrument.

Reference has already been made to the original form of receiving apparatus called the Coherer. This, however, has been very largely superseded by the Magnetic Detector of Marconi and the Crystal Detector, both of which make the signals perceivable as buzzing sounds in the telephone.

The magnetic detector owes its existence to the fact that oscillations tend to destroy magnetism in iron. It is believed that every molecule of iron is itself a tiny magnet. If that be so one would expect every piece of iron to be a magnet, which we know it is not. We can always make a piece of iron into a magnet by putting another magnet near it, but when we take the other magnet away the iron loses its power, or to be precise it *almost* loses it. A piece of even the best and softest iron having once been magnetized retains a little magnetic power which we call residual magnetism.

All this is easily explained if we remember first that a heap of tiny magnets lying higgledy-piggledy would in fact exhibit no magnetic power outside the heap. If, however, we brought a powerful magnet near them it would have the effect of pulling a lot of them into the same position, of arranging them in fact so that instead of all more or less neutralizing each other they could act together and help each other. Then the heap would become magnetic. On removing the powerful magnet, however, a lot of the little ones would be sure to fall down again into their old places and so the heap would at once lose a large part of its power, yet some would remain and so it would retain a certain amount of residual magnetism. If, then, you were to give the table on which the little magnets rest a good shake, the higgledy-piggledyness would be restored and even the residual magnetism would vanish.

So we believe that the little molecules lie just anyhow, wherefore they neutralize each other and the mass of iron is powerless. When another magnet comes near, however, they are more or less pulled into the right position and the iron becomes magnetized. When the magnet is removed the magnetism which it produced is largely lost, and if last of all we give the iron a smart blow with a hammer even the residual magnetism vanishes too.

Now, oscillations taking place in the neighbourhood of a piece of iron possessing residual magnetism have much the same effect as the blow of a hammer. Probably because of its rapidity an oscillating current shakes the molecules up and strews them about at random, entirely destroying any orderly arrangement of them. And Marconi used that fact in detecting oscillations.

Two little coils of wire are wound together, one inside the other. Through the centre of the innermost there runs an endless band of soft iron wire. Stretched on two rollers this band travels steadily along, the motive power being clockwork, so that it is always entering the coil at one end and leaving it at the other. As it travels it passes close to two powerful steel magnets, so that as it enters the coil it is always slightly magnetized. The oscillations are passed through one of the two concentric coils, and their action is to remove suddenly the residual magnetism in that part of the moving wire which is at the moment passing through. That sudden demagnetization then affects the second of the concentric coils, inducing currents in it, not of an oscillating nature but of an ordinary intermittent kind which can make themselves audible in a telephone which is connected with the coil. This arrangement, then, causes the oscillations, which will not operate a telephone, to produce other currents of a different nature which will.

The reason why oscillations have no effect in a telephone is no doubt because they change so rapidly, at rates, as has been mentioned already, of the order of a million per second. The telephone diaphragm, light and delicate though it is, is far too gross and heavy to respond to such rapidly changing impulses as that. In the magnetic detector the difficulty is overcome by making them change the magnetic condition of some iron wire which change in turn produces currents capable of operating a telephone. The Crystal Detector achieves the same result in another way.

There are certain substances, of which *carborundum* is a notable example, which conduct electricity more readily in one direction than the other. Most of these substances are crystalline in their nature, and hence the detector in which they are used gets its name. *Carborundum*, by the way, is a sort of artificial diamond produced in the electric furnace and largely used as a grinding material in place of emery.

It is easy to see that by passing an oscillating current, which is a very rapidly alternating current, through one of these one-direction conductors one half of each oscillation is more or less stopped. Oscillations, again, are surgings to and fro: the crystal tends to let the *tos* go through and to stop the *fros*. That does not quite explain all that happens. It is not fully understood. The fact remains, however, that by putting a crystal in series with the telephone the oscillations become directly audible. The term *in series with* means that both crystal and telephone are inserted in the local receiving circuit so that the currents in that circuit pass through both in succession.

The resistance of the crystal being very great, a special telephone is needed for use with it. It is quite an ordinary telephone, however, except in that it is wound with a great many turns of very fine wire and is therefore called a high-resistance telephone.

Whichever of these detectors be used, then, the operator sits, with his telephone clipped on to his head, and with his tuner set for that wave length at which his station is scheduled to work, listening for signals. He may go for hours without being called up, and in the meantime he may hear many signals intended for others. He knows they are not for him, since every message is preceded by a code signal indicating to whom it is addressed.

Under the conditions of warfare there is far more listening than there is sending, but when a station wishes to send the operator just switches over, cutting out his receiving apparatus and bringing his transmitting instruments into operation, and, having adjusted his tuner for the wave length of the station to which he desires to communicate, he flings out his message.

In war-time, too, there is much listening for the signals of the enemy, which is the reason why as few messages are sent out as possible. In this case the man sits with his telephone on his head carefully changing his tuner from time to time in the endeavour to catch any message in any wave-length which may be travelling about. This searching the ether for a chance message of the enemy must be at times a very wearisome job, but it must be varied with very exciting intervals.

On aircraft it is clear that no earth connection is possible. The antenna in that case usually hangs vertically down from the machine or airship. Under these conditions the valuable effect of the earth connection is of course lost. As will be remembered, the earth-connected apparatus sends forth waves which cling more or less to the neighbourhood of the earth's surface, while those from the non-earthed apparatus as used by aircraft tend to fly in all directions. The latter apparatus is in fact almost precisely similar to that which Hertz used in his first experiments. Hence the range is comparatively poor under these conditions, but it is good enough for very valuable work in warfare. Communication between airman and artillery by this means has revolutionized the handling of large guns in the field.

To save the airman from the accidental catching of his aerial wire in a tree or on a building there is sometimes fitted a contrivance of the nature of wire-cutters so that he can at any moment cut himself free from it.

So far we have dealt almost exclusively with the naval and aerial use of this wonderful invention. It is employed, though in a lesser degree, in land warfare. In such cases the aerial may be merely a wire thrown on to and caught up on a high tree. More elaborate devices are used, however, such as a high telescopic tower similar to the tall fire-escape ladders of the fire-brigades. Anyone who has seen the ladders rush up to a burning building and commence to erect themselves almost before they have stopped will realise how valuable such a machine must be for forming a temporary and easily movable wireless antenna. The power which causes the tall tower to extend itself erect in a few seconds is compressed air carried in cylinders upon the machine, while the power which takes it from place to place is a petrol motor, and since the latter can be made to re-charge the storage cylinders it is clear that in it we have a marvellously convenient adjunct to the wireless apparatus.

But apart from such carefully prepared devices the men of the Royal Engineers are past masters in the art of rigging up, according to the conditions of the moment, all sorts of makeshift apparatus whereby signalling over quite long ranges can be carried on by *wireless*. Such improvisations, could they be recorded, would constitute war inventions of a high order.

CHAPTER 22

Military Telegraphy

Telegraphy plays a very important part in warfare. The commander of even a small unit cannot see all that his men are doing or suffering, but is kept posted by telegraph or telephone, while communication between units depends very largely indeed upon such means. Wireless telegraphy, in land warfare, is largely devoted to communication between aircraft and the artillery batteries with which they are working, and to avoid interference with that important work telegraphy *by wire* is employed for most other purposes.

Right at the front this communication is kept up by means of that type of instrument which the soldiers call a *buzzer*, for the good and sufficient reason that that is really what it does.

In view of the fact that soldiers speak of their home-land, for which they are enduring all manner of risk and hardship, and to which they are longing to return, by the contemptuous-sounding name of Blighty, we might expect that what they call a buzzer has nothing whatever to do with making sound, but in this case the name describes the thing very aptly. Its sole purpose and intent is to make buzzing sounds of either long or short duration.

Perhaps the simplest way in which I can describe this useful and interesting invention is by telling you how you can make one for yourself. It is nothing more than an electric-bell mechanism connected up in a certain way.

As most people know, an electric bell contains a magnet made of two round pieces of iron placed parallel and yoked together at one end by means of a third piece of iron, generally flat, while on to each round piece is threaded a bobbin of insulated wire. The iron becomes a magnet when, and only when, current flows through the wire.

Near the free ends of the round pieces, or the poles of the magnet, to use the orthodox term, is placed another little piece of iron called the armature, carried upon a light spring. When the current flows in the wire the armature is pulled towards the poles against the force of the spring, but when the current ceases the magnet lets go and the armature, urged by the spring, swings back again.

Behind the armature is a little post through which passes a screw tipped with platinum, and in operation this screw is advanced until its point touches a small plate of platinum carried by the armature. Connection for the current is made to this contact screw whence it passes to the armature, through the spring to the wire upon the magnet, through that and away. On completing the circuit, then, as when you push the button at the front door, current flows and energizes the magnet. A moment later, however, the armature moves, breaks the contact with the screw and stops the current. Then the magnet lets go and the armature springs back, making contact once more and setting the current flowing again. These actions repeat themselves over and over again quite automatically, and the hammer which is attached to the armature vibrates accordingly.

That is the ordinary familiar electric bell. Cut off the hammer and you have a buzzer with which excellent telegraph signals can be sent.

So much for the sending apparatus. The receiving device is simply an ordinary telephone receiver. There is sometimes a little confusion in people's minds because of this. A telephone is used, but it is used as a telegraph instrument. The sounds heard in it are not speech but long and short buzzing sounds which, being interpreted according to the code of Morse, deliver up their message.

Now the telephone, by which term is always meant the receiver (the sending part of the telephone apparatus being a *microphone*), is one of the most remarkable pieces of electrical apparatus which the mind of man has ever conceived. It is astonishingly robust. With ordinary care you cannot damage it. There is no need whatever to keep it wrapped in cotton wool or even to keep it in a case. Without harm you can put it loose in your pocket. Within reason you may even drop it a few times without harm. Its cost is only a few shillings. Yet its sensitiveness is simply astounding. It will detect the existence of currents so small that any other type of instrument to deal with them has to be extremely delicate and costly.

It consists of a magnet fitted into a little brass case with a little piece of soft iron fixed on each pole, while each of these *pole-pieces* is surrounded by a tiny coil of wire. The lid of the box is a disc of thin sheet-iron, and things are so proportioned that the pole pieces nearly but not quite touch this sheet-iron *diaphragm*.

An outer cover, generally of ebonite, serves to catch the sound-waves caused by any movement of the diaphragm and convey them to the ear.

The action of the permanent magnet tends to pull the diaphragm inwards—to bulge it in slightly—so that it is in a state of very unstable equilibrium. Because of this instability a very tiny current flowing through the coils and either adding to or subtracting from the strength of the magnet is sufficient either to draw it still closer or to let it recede a little. Whether it approaches or recedes depends upon the direction of the current through the coils and makes no difference to the sound. The movement of the diaphragm is great or small according as the current is strong or weak: any variation in the current causes a perfectly corresponding movement in the diaphragm. Even those very small and very complex changes in air-pressure which give us the sensation of sound are very faithfully followed by this simple bit of sheet iron, so that the sounds are faithfully reproduced for our benefit. At the moment, however, we are not dealing with speech but with buzzing sounds, which are very simple, being merely a rapid succession of *ticks*.

The telephone, it must be remembered, takes no notice of a steady current, except when it starts and stops. But each time that occurs it gives a tick. Hence, if we start and stop a current very rapidly, or to use another term, make it rapidly intermittent, we get a rapid succession of ticks, and if rapid enough they form a humming, buzzing, or singing sound. If very fast you can get a positive shriek. The precise character of the sound depends entirely upon the rapidity of the intermittency.

Now it is easy to see that the current passed through an electric-bell mechanism is intermittent. It is the very nature of the apparatus to make the current intermittent. It is by so doing that it works. Therefore, if we pass the same current which works a bell through a telephone we get a buzzing or humming sound according to the speed of interruption.

The vibration of the armature itself also causes a humming sound of a similar note or tone to that heard in the telephone, but it must be clearly understood that these two sounds are quite different. One is the result of mechanical motion, the other is the result of electrical action producing motion in the diaphragm of the telephone. When you listen in the telephone it is not that you hear the sound of the bell mechanism, you hear another sound altogether, although, since both have the same origin, both have the same note or tone.

Take any old bell, then, which you may happen to have or be able to procure and an old telephone such as can be bought for a shilling or so at a second-hand shop, and these together with a pocket-lamp battery can be formed into a military field telegraph.

The way to connect these up is to run a wire from one of the copper strips on the battery to one of the terminal screws on the bell, a second wire from the other screw on the bell to one of the flexible wires of the telephone, which may be a mile away if you like, a third wire returning from the other flexible wire of the telephone back to the battery. To send signals all you have to do is to touch the return wire upon the second strip of the battery for short or long intervals, thereby making the dot-and-dash signals. Or a simple form of key can easily be contrived for the purpose.

Every time you complete the circuit the buzzer will buzz, in other words, it will permit an intermittent current to pass round the circuit and a buzzing or humming sound will be heard in the telephone, no matter how far away it may be.

This arrangement, however, involves two wires between the two stations, and in practice only one is usual. This could be arranged by running the third wire from the telephone not back to the sending station but to a peg driven into the earth, connecting the second pole of the battery in like manner to an earth pin at the sending end. Thus the return wire would be done away with and the earth utilized instead. To do that, unfortunately, you would need to increase very greatly the power of your battery, for although the path through the earth itself offers practically no resistance at all to the current, the actual places where the current passes to earth and from earth, especially if they be simply temporary pegs driven into the ground, offer very considerable resistance, so that in order to get enough current through the buzzer to make it work would need a powerful battery. There is another way, however, by which that difficulty can be overcome quite easily.

Probably all my readers know something of the induction or shocking coil, wherein intermittent currents in one part of the coil induce intermittent currents of a somewhat different kind in another part of the coil. Few people realize, however, that the same effect can be attained, within limits, in a single coil such as the winding upon the magnet of an electric bell.

Watch a bell at work and you will notice a bright spark at the place where the contact is made and broken. That spark is due to a sudden rush of current which takes place in the coil when the original current is stopped, in other words, when the contact is broken. It is as if the coil gives a rather vicious kick every time the current is stopped. There is not much electricity in this kick current, but it is very forceful, and it is that force which makes it actually jump across the gap after contact has been broken, thereby causing the spark.

Now we can capture most of that energy and make it go a long distance through wire and through earth carrying our messages for us. To do this we need to make a new connection on the bell at the place where the spring is fixed. Then we can make two circuits. One is between the two terminal screws of the buzzer, in which circuit we must include the battery and the key. That circuit will be just as it would be if we were fixing the buzzer to announce our visitors at the front door.

The second circuit is different: lead one wire from the new connection just made and take it to a pin driven into the ground. If the ground is just a shade moist a wire meat-skewer will answer admirably. Then lead a second wire from that one of the two terminal screws which is connected directly to the winding of the magnet (not to that one which is connected to the contact screw) and lead it away to your distant station.

At the other station connect the single wire to the telephone as before and the other end of the telephone to a pin in the earth. You will find that the kicks from the coil will traverse wire and earth-return quite easily, while there will be no difficulty about working the bell, for the small battery will do that quite well. In fact, after cutting the hammer off and so converting a bell into a buzzer, I have got quite good results with one-third of a pocket-lamp battery. The little flat batteries so familiar to us all if divested of their outer covering will be found to consist of three little dry cells any one of which is quite capable of sending messages in the way described as far as any amateur is likely to want to send.

To be able to send and receive at either end it is only necessary to connect both telephones and both coils *in series*. That is to say, connect one end of the coil to the long wire and the other to one wire of the telephone, the other wire of the telephone being connected to earth. If this be done at both ends signals can be sent and received both ways.

Many young readers, scouts, members of cadet corps and the like, will find great pleasure and interest in constructing and working this apparatus, besides which it shows precisely what the official *buzzer* is like.

Although beautifully made, of course, the army instrument is essentially just that and little more. It has an additional feature, however, namely, a microphone, so that when desired it can be used as a speaking telephone for transmitting verbal messages. It also has the bottom of the case made of a brass plate so that earth pins are often unnecessary, the case dumped down upon the ground being a good enough *earth*.

Buzzers are not used for very long lines: forty miles is about the limit, and usually the distances are very much less. That is because long lines rather object to rapidly changing currents flowing through them. Why, you say, what currents could change more rapidly than telephone currents carrying speech, yet they go for hundreds of miles? True, but in that case there are two wires, flow and return, twisted together all the way, under which conditions they interact upon each other in such a manner as to abolish the difficulty to which I am referring. Buzzers and indeed all the telegraph circuits consist of one wire and the earth, which is quite different.

Another objection to the buzzer is that it is apt to interfere with others. For instance, if two buzzer sets are at work anywhere near each other and the wires run parallel for a distance they will be able to hear each other's signals as well as their own. If two such sets are earthed near together the same thing happens, the signals of one are picked up by the other, a very annoying state of affairs for the operators.

Right at the front, however, amid the rough and tumble of the actual fighting, the buzzer is supreme. The wire used is sometimes plain copper enamelled: more often, however, it is a mixture of steel and copper strands twisted together and covered with a strong insulating covering. This is carried on reels in properly fitted carts which can advance at a gallop, paying out the wire as they go. The inner end of the wire is connected to the axle of the reel in such a way that a telegraphist in the cart is in communication all the time with the starting-point, the wheels of the cart providing him with an earth connection.

When laying these wires another interesting little device is often used—an earth plate on the operator's heel. Thus, while carrying the wire along, laying it as he goes, he can still be in communication with the starting-point every time he puts his heel to the ground.

For the longer lines away back from the fighting the methods employed are just the same as those of peace. *Sounder* instruments are used, Wheatstone automatic machines, duplex and quadruplex systems, whereby two and four messages are sent simultaneously over the same wire, indeed all the contrivances and refinements of the home telegraph office are to be found in the field telegraph offices. But it would hardly be fitting to describe them here. Some information on the subject will be found in *The Romance of Submarine Engineering*, where their application to cable telegraphy is dealt with.

A genuine speciality of warfare, however, is the methods by which makeshift arrangements can be set up, such as sending telegraph messages over a telephone wire without interfering with the latter.

Imagine that A and B are the two wires of a telephone circuit running (for the sake of simplicity) from north to south. At the south end I connect a telegraph set to both wires while you, we will imagine, do the same at the north end. You and I can then signal to each other without the telephone man hearing us at all. To him the two wires are flow and return, to us they are both *flow*, the *earth* being our return. Thus our signals never reach his instruments at all. But when we each connect to both his wires, do we not *short-circuit* or connect them to each other, thereby destroying his circuit? No, we are too cunning for that. We first connect the two wires A and B together with a coil of closely wound wire, having, in scientific language, much *inductance*, and telephone currents shun a coil of that sort. Then we make our connection to the centre of that coil so that our currents go to A through half the coil and to B through the other half. This enables us to use the apparatus without interfering with the other fellow at all. For this, by the way, we must use ordinary telegraph instruments. We cannot employ a buzzer, for these coils which we use to obstruct the passage of the other man's telephone currents would also obstruct the changing currents from a buzzer. The slow, steady currents of the ordinary telegraph pass quite easily, however.

Again, suppose you and I want to communicate by buzzer and there is already a wire laid passing both of us but in use already for ordinary telegraphy. We only need to add a *condenser* to our apparatus and we can manage all right. As a matter of fact, the service instruments generally have condensers partly for this very purpose. Each of us then connects his instrument to the wire and to earth, after which we can signal to each other while the telegraphist is unaware of the fact. The reason that is possible is the reverse of what we saw just now. There we had a coil which obstructed buzzer or telephone currents but passed ordinary telegraph currents. Here we use condensers which will pass our buzzer currents but not the ordinary telegraph currents.

Thus the soldier telegraphist is up to many dodges whereby he can save time or save material, both of which may be precious. As in bridge building and other branches, he needs to be quick to adapt himself to circumstances, to utilize to the full any opportunities which may present themselves. But his principles are quite simple and do not differ in any way from those of peace. It is only in applying them that the differences arise.

CHAPTER 23

How War Inventions Grow

The inventor of one of the devices described later on in this book modestly claims that he did not invent it but it invented itself. What he means is that he worked step by step, from simple beginnings, each step when complete suggesting the next. To put it another way, many inventions grow in the inventor's mind, sometimes from unpromising beginnings, the most unlikely start often resulting in the most successful ending.

Who has not heard of the *tanks* which made such a name for themselves when they suddenly appeared in Northern France? The British commander-in-chief simply mentioned that a new type of armoured car had come into use with good results, but the newspaper men set the whole non-Teutonic world laughing with droll stories of huge monsters suggestive of prehistoric animals which suddenly began to crawl through the slime and mud of the battle-field, pouring death and destruction upon the astounded Germans.

How they came to be called tanks no one seems to know clearly but that is how they will be known for all time. It has been suggested that they were so named because tank is one of the things which they certainly are not, the intention being thereby to add to the mystification of the enemy. That is by the way, however, for we are more concerned with the things than with their name.

Their precise origin is wrapped in mystery but we have it on excellent authority that they grew out of the peaceful tractor, originally intended to drag a plough to and fro across a field in the service of the farmer. An illustration of one of these interesting machines will be seen in this book which will well repay a little study.

It consists of a steel frame or platform upon which is mounted a four-cylinder petrol engine with a reservoir above to carry the supply of fuel and with a radiator in front to cool the water which keeps the engine from becoming too hot. Towards the back of the vehicle is what is called by engineers a worm-gear, the function of which is to reduce the one thousand revolutions per minute of the engine to somewhere near the slow speed required of the wheels of the tractor.

This worm-gear is simply a wheel with suitable teeth on its edge in conjunction with a screw so made that its thread can engage comfortably with the teeth. This latter, because of the wriggling appearance which it presents when it is revolving is called a worm, which name it gives to the whole apparatus. Both wheel and worm are mounted in bearings which form part of a case enclosing the whole so that dirt is excluded while, the case being filled with oil, ample lubrication is assured. The shafts of both wheel and worm emerge through holes in the case.

It will easily be seen that each single turn of the worm will propel the wheel one tooth, so that if the wheel have fifty teeth, for example, the worm will turn fifty times to the wheel's once. Thus a great reduction in speed is attainable with this device and what is equally valuable, a great increase of power also results. Thus a small engine, working at a high speed, is able by means such as this to pull very heavy loads at a slow speed.

It is evident, however, that the reduction necessary in this case cannot be attained even by a worm-gear, for there are other wheels visible which show that ordinary tooth gearing is also employed to reduce the speed even further before it is applied to driving the tractor along. Practically all the other gear which we see in the picture, above the platform, consists of the controlling apparatus.

The object with a screw-like appearance just behind the engine is not really a screw but is a flexible coupling joining the engine to the worm-gear, its flexibility enabling the two to work sweetly together even though by chance they may get just a little out of line with each other.

But by far the most interesting part of the machine is that which is underneath the frame. At one end we see a pair of ordinary-looking wheels and between them the gear for swinging them to right or left for steering purposes, but even they are somewhat unusual, since they will be seen to have flanges or rims round the edge for the purpose of biting into the earth, so that they may be able to guide the machine the better in soft ground.

The back wheels, however, are quite peculiar, for there is a pair on each side and round each pair is a chain somewhat after the fashion of a huge bicycle chain. The links of this chain are made of tough steel and they are two feet wide, so that each chain forms a broad track upon which the machine moves. The links of this track-chain will be seen to be tooth-shaped so that they grip or bite deeply into the yielding ground. The teeth, moreover, are shaped like those of a saw and they are so placed as best to help the tractor forward.

Between the two chain-wheels will be noticed a row of smaller wheels and it is these which largely support the weight of the machine, the chains forming tracks upon which they run.

The wheels actually turned by the power of the engine are the chain-wheels, and their action is such as to keep on laying down and then taking up again two broad firm tracks along which, at the same time, they keep propelling the other wheels which carry the weight above. The effect, really, is just as if the machine had a pair of driving wheels two feet wide and of enormous diameter, of such diameter, in fact, that the part in contact with the ground is almost flat. Thus there is always a broad bearing surface to prevent sinking in soft earth, while the tooth-like shape of the links gives a firm hold even under very adverse conditions.

This form of construction has been used for some few years now under the name of *caterpillar* or *centipede* traction. A glance at the picture will explain those names, particularly if the chain-driven part of the vehicle be imagined to be a little longer than it is in the particular machine shown.

The idea of armouring a vehicle with bullet-proof plates is also a fairly old conception. Armoured trains were used again and again during the South African War, and armoured motor-cars became familiar to most people. In the case of cars, however, the armour could only be very light and the guns carried were limited practically to a single machine-gun and some rifles. Moreover, the operations of a car are very largely confined to such places as are blessed with good roads or smooth plains. An armoured car of the older type would have cut a poor figure amid the shell-holes and mine-craters of Northern France. It would have had to keep to the roads and so it was little used.

But the idea of an armoured vehicle was good and a good idea is never entirely lost. Sooner or later some genius puts it to good use. Thus the idea of an armoured vehicle came to be associated with the idea represented in the centipede tractor and the result was the tank.

Why not armour a large centipede, said someone? Make it very big and strong. It will trample down the barb-wire entanglements as if they were grass. If made long enough and rightly balanced it will pass over the trenches like a moving bridge. Nothing but a direct hit from

254

THE PARENT OF THE TANK

HERE WE SEE AN INNOCENT AGRICULTURAL TRACTOR WITH CATERPILLAR HIND WHEELS. IT IS OUT OF SUCH A MACHINE THAT THE IDEA OF THE FORMIDABLE TANK WAS EVOLVED.

a heavy gun will do it much harm. For, observe, the mechanism can be entirely covered up, all the vital parts can be well protected, and the chain tracks can be so strong as to be almost undamageable.

Thus we get a glimpse of the growth of this simple peaceful agricultural machine into one of the most striking mechanical achievements of the Great War.

Another thing which seems to have grown more or less of itself is the bomb or grenade. Before the time of modern accurate fire-arms hand-grenades were quite a recognized weapon. The Grenadier Guards owe their title to this fact and carry the design of a bursting grenade upon their uniforms. Yet until a few years ago everyone thought that such things were done with for ever: that with modern rifles soldiers would seldom get near enough together to use grenades and that if they did the bayonet would be the weapon to be used.

When, however, the Germans were driven back at the battle of the Marne and found themselves compelled to entrench in order to avoid further disaster, it soon became evident that neither rifle nor bayonet nor both together entirely filled the needs of the infantryman.

Since the Allies were not powerful enough to drive the Germans from their trenches forthwith, they, too, had to entrench. Gradually the trenches drew nearer and nearer together and at the same time skill in entrenching increased. Thus a time soon arrived when both rifle and bayonet were largely useless for purposes of offence. Then the hand-grenade came into its own again, for the men could throw it from the depths of their own trench high into the air in the hope that it would fall into the trenches of the enemy. The call for these quickly produced the supply. There is little need to describe them here, for who among us has not intimate friends who used them again and again? This much may be said, however. They were little hollow balls of cast iron, sometimes chequered so that when they burst they flew into many fragments. Inside was a charge of explosive with a suitable fuse or firing mechanism. Some were fixed to the end of a stick for convenience in throwing, while others were simply handled like a cricket-ball.

They serve to show us, however, how an old idea may under fresh conditions be revived into what is practically a new invention.

Another example of the same sort is the revival of chain mail. Who, but a few years ago, would have thought it possible that modern soldiers would go to battle sheathed in shirts consisting of little metal plates cunningly connected by wire links and so overlapping each other as to form a perfect shield for all the more vital parts of the body? To what extent these were worn I do not know, for the British soldier is a very shy fellow in some ways and there are few who would not be a trifle ashamed to let their comrades see them thus garbed. They would feel that it was a confession of fear, and however afraid an Englishman may be he will never admit it. He is really a pious fraud, for the more he is really afraid inwardly the more courageously will he act just to hide his fear.

Since, however, the bullet-proof helmet is worn officially nowadays there seems no reason whatever why the bullet-proof waistcoat should not be adopted officially too. It is very light and very flexible and it is claimed that it is quite effectual in stopping rifle and machine-gun bullets.

Thus we see in what different ways inventions grow. Some are warlike from first to last, like the gun and the torpedo, but we find a vast range of peaceful things growing into implements of warfare, as the farmer's tractor has been developed into the tank, while not less interesting are the old ideas revived and adapted to modern needs, exemplified by the hand-grenade and the chain armour.

CHAPTER 24

Aeroplanes

Of all the great inventions perhaps the most striking because of the suddenness with which they have come upon us are those relating to the navigation of the air. Until a few years ago to fly was taken to typify the impossible. Now we see men flying every day and there is scarcely anyone who has not had a friend or relative in the Flying Corps.

Recent experience, too, has shown that this one invention has revolutionized warfare in several important departments, particularly in the use of very heavy long-range artillery. Huge guns, hidden in a hollow or behind a hill, have been set to throw shells on to an unseen target, while a man in an aeroplane above watches the result and signals back by wireless. Thus by the aid of aircraft the power of artillery has been immensely increased.

Again, aircraft have superseded cavalry for reconnaissance purposes, that is to say, for finding out the enemy's strength and preparedness. Only a few years ago a General who needed information as to his foe would send forward a screen of cavalrymen who would cautiously creep forward until, judging by what they could see and by what sort of a reception they got, they were able to form some idea of the foe's arrangements. Nowadays, however, the airmen sail over his head and take photographs of him and his positions. A careful commander to-day not only screens his men and his guns from view along the land but he also tries his best to make them invisible from above. And, speaking of inventions, the soldiers have shown a degree of ingenuity in making themselves and their guns invisible which almost merits a volume to itself.

The airman, therefore, goes up and sails over the enemy. He may be simply observing for some particular unit of artillery, or he may be sent to find out things generally—nothing in particular, but anything which seems likely to be of use. He looks out intently and carefully, moreover he not only looks with his own eyes: as has just been mentioned, he takes photographs, which can be developed on his return and studied minutely at leisure. He may, or may not, according to circumstances, send back reports of an urgent nature by wireless telegraphy.

In some cases these duties are all carried out by one man, but in others there are two: one the pilot who looks after the working of the machine, and the other the observer whose whole attention can thus be devoted to scrutinizing the enemy.

Of course, when aeroplanes go on scouting expeditions like this they are apt to be attacked by the enemy both by anti-aircraft guns and also by other aeroplanes. The former can only be met by high speed and the steering of a somewhat erratic course so as to confuse the gunners and prevent them from taking good aim.

The other aeroplanes, however, must be met by actual fighting. The only way to defeat them is to go for them and attack them, a machine-gun being the most usual weapon.

Besides those who go up for definite scouting operations or to *spot*, as it is termed, for the artillery, there are other machines whose sole duty is fighting. These go up for the purpose of driving off those machines of the enemy which may come prying, or to keep the ground, so to speak, for the scouting machines and enable them to do their work unmolested.

Then there are, of course, still others whose function is to carry out bombing expeditions.

All these different duties call for different types of machine, but I do not propose to go into the differences here since changes are so rapid in this particular field that only the general principles remain unchanged for any length of time. What has just been hinted, however, as to the different kinds of work which the aeroplane is called upon to do will enable the reader to see why different kinds of machines are needed.

So far we have only spoken of aeroplanes. There is a kind of machine sometimes called a hydroplane but which we are gradually getting to call a sea-plane. The latter term is much to be preferred, since the former is also in use to denote a special kind of high-speed boat.

Now a sea-plane only differs from an aeroplane in that it has floats instead of wheels. The aeroplane has wheels to enable it to alight upon and arise from the ground: the sea-plane has floats by which it can alight upon the water and arise from the water also.

In some instances this float idea is made so pronounced a feature of the machine that it becomes a flying boat.

Sea-planes are therefore really only aeroplanes specially adapted for a certain purpose. They are really just as much aeroplanes as those machines which go by that name. It is somewhat unfortunate, therefore, that a separate term is used to describe them. But there it is: names grow in a very curious way, not always in a logical way, and a name having once stuck to a thing in the mind of the public it is very difficult to make any alteration.

Aeroplanes, then, may be said to include a subdivision known as sea-planes, and for the rest of this chapter what is said of aeroplanes will apply to sea-planes also.

Without doubt, these are the fastest vehicles in existence. Many of them can exceed a speed of a hundred miles an hour. Consequently, the pilot lives while he is aloft in the equivalent of a furious gale, and it would seem as if that must produce such a degree of cold as to be almost unendurable. Moreover, it appears that this cold is almost as bad in summer as in winter, for the temperature high up in the air is much the same all the year round. The consequent muffling up with thick clothes and gloves, while it mitigates the cold, must add greatly to the pilot's difficulties in managing his machine. The protection for his eyes and ears which is made necessary by the same conditions must likewise add to his difficulties or at any rate to his discomfort. On the other hand, the effect of gliding at a very high speed over a perfectly smooth track, for that is in effect what it is, is very exhilarating, which to some extent compensates for the other drawbacks.

Moreover, the handling of such a machine in the air, particularly if a fight is included in the programme, appeals strongly to the sporting instincts of young men, so much so that during the War, in spite of the dangers and hardships, and the continual loss of life, there was never a dearth of men anxious to become pilots.

Owing to these considerations, too, it follows that the best aviators are to be found in those lands where the people are most devoted to sports. Hence, as we have it on excellent authority, the young men of Great Britain and the United States, with their love of adventure and their strong sporting instincts, make better men in the air than the Germans.

But really we are more concerned here with the machines than with the men, so let us get back to our subject.

The aeroplane consists of one or more *planes* or surfaces which, on being held at a certain slant and then pushed forward rise or remain supported in the air. Therefore the plane or planes need to be supplemented by first a tail and horizontal rudder to hold them at the correct slant, and an engine and propeller to drive them forward.

It is not necessary, here, to go over the history of the aeroplane, as that has been told so often. It is not of much interest, moreover, except to those who are particularly concerned with small details of construction, for in a general way the machine of to-day is very little different from one pictured by Sir George Cayley a hundred years ago. It is only the perfecting of the details which has transformed a dream into a very real thing.

So we will look only at the construction of the aeroplane in a general way, to do which we must first consider why it flies at all. It is due to the well-established law that action is always accompanied by a reaction equally strong and in the opposite direction. When a gun is fired the explosion not only drives the shell forward but equally drives the gun itself backward. The backward energy of the recoil is precisely equal to the forward energy of the shell. The two are equal but in opposite directions. In like manner a rocket ascends because the hot gases from the paper cylinder blow forcibly downwards, thereby producing an equal reaction upwards.

Now the plane of a flying machine is held with its forward edge a little higher than its rear edge, so that as it is pushed along it tends to catch the air and throw it downwards. Hence the reaction tends to lift the plane upwards. When the machine starts the reaction is not sufficient to overcome gravity, which is trying to hold the machine down upon the ground, but as the speed increases and the air is thrust down with more and more violence the point is ultimately reached when the reaction is able to overcome gravity and the machine ascends.

When a sufficient height is reached, the pilot alters the position of his horizontal rudder or *elevator* so as to make the position of the plane more flat, with the result that it throws the air downwards to a less extent, and the reaction is thereby reduced until it is only just sufficient to keep the machine at the same height. To descend, the position of the plane is made still flatter, the reaction is reduced still more and gravity has its way once again, bringing the machine to earth.

In other words, the machine acts under the influence of two forces: the downward pull of gravity and the upward reaction due to the action of the machine in throwing the air downward. The former never varies, the latter can be varied by the pilot at will: he can increase it by increasing the speed or by increasing the tilt of his plane or planes: he can reduce it by diminishing the speed or the tilt. Since generally speaking the speed of his engine will remain constant, he rises, remains at the same height or falls, at will, by the simple manipulation of the elevator through which he can change the tilt or inclination.

Most machines have a fixed tail as well as a horizontal rudder or elevator, the same being so set that it tends to keep the plane in a certain normal inclination, the elevator being called in to increase that or diminish it as may be required.

In addition to the elevator there is also another rudder of the ordinary kind, such as every ship and boat has, for guiding the machine to right or left. The elevator steers up and down, the rudder steers to either hand.

Provision is also made for balancing the machine. This is sometimes in the form of two small planes hinged to the main plane, one at either end, connected together and to a controlling lever by wires, so that by their use the pilot can steer the right-hand side of his machine upwards and the left-hand downward, or vice versa, if through any cause he finds a tendency to capsize.

In some machines the same effect is produced not by separate planes but by pulling the main plane itself somewhat out of shape, but precisely the same principle is involved.

The planes are usually made with a slight curve in them, so that they may the better catch the air and *scoop* it downwards, so to speak. They usually consist of fabric specially made for the purpose, stretched upon a light wooden framework. The whole framework is usually of wood with metal fittings frequently made of aluminium for the sake of lightness.

The engines have been mentioned in another chapter. The propeller which is almost invariably fixed directly upon the shaft of the engine has two blades only and not three as is usual with those of ships. Precisely why this should be so is not clear, but experience shows that two-bladed propellers are preferable for this work. They are made of wood, several layers being glued together under pressure, the resulting log being then carved out to the required shape. This makes a stronger thing than it would be if cut out of a single piece of wood.

All parts, engine, elevator, rudder and balancing arrangement, are controlled by very simple means from the pilot's seat.

In monoplanes there is but one main plane, resembling a pair of bird's wings. Or if we care to look upon it as two planes, one each side of the body, then we must call it a pair. Since the name *mono* indicates one it is best to think of it as one plane although it may be in two parts. The biplane has, as its name implies, two planes, but in that case there can be no doubt, since they are placed one above the other. Machines have been made with three planes and even with as many as five, but monoplanes and biplanes appear to hold the field.

It is not possible for an aeroplane to be in any sense armoured for protection against bullets: for defence the pilot has to depend upon his own cunning manoeuvres combined with the fast speed at which he can move. For offensive purposes he usually has a machine gun mounted right in front of him with which he can pour a stream of bullets into an opponent or even, by flying low, he can attack a body of infantry. It is recorded that one German prisoner during the war, speaking of the daring of the British pilots in thus attacking men on foot, exclaimed, *They will pull the caps off our heads next.*

Some of the aeroplanes have their propeller behind the pilot and some have it in front. The latter, to distinguish them, are called *tractor* machines, since in their case the propeller pulls them along. Now it is easy to see that a difficulty arises in such cases through the best position for the gun being such that it throws its bullets right on to the propeller. But that has been overcome in a most simple yet ingenious way. The gun is itself operated by the engine with the result that a bullet can only be shot forth during those intervals when neither blade of the propeller is in the way. The propeller is moving so fast that it cannot be seen and the bullets are flying out in a continuous rattle, yet every bullet passes between the blades and not one ever touches.

It is easy to see that when an aeroplane is manned by a single man, as is often the case, he must have his hands very full indeed, what with the machine itself and the gun as well. In fact, he often has to leave the machine for a short time to look after itself while he busies himself with the gun.

Now there we see a sign of the wonderful work which has been done in the course of but a few years in the perfecting of the aeroplane, the result of a series of improvements in detail which make but a dreary story if related but which make all the difference between the risky, uncertain machine of a few years ago and the safe, reliable machine of to-day. Modern machines are inherently stable. The older ones had the elements of stability in them but they were so crudely proportioned that these inherent qualities did not have a chance to come into play.

If one drops a flat card edgewise from a height it seems as if it ought to fall straight down to the ground. Yet we all know from experience that it seldom does anything of the kind. Instead, it assumes a position somewhere near horizontal and then descends in a series of swoops from side to side. There we see the principle at work which, in a well-designed aeroplane, causes inherent stability. The explanation is as follows.

The aeroplane is sustained in the air through the upward pressure of the air resisting the downward pull of gravity. That has been fully explained already. Now gravity, as we all know, acts upon every part of a body whether it be an aeroplane or anything else. But for practical purposes, we may regard its action as concentrated at one particular point in that body, called the centre of gravity. Likewise, the upward pressure of the air acts upon the whole of the under surface of the plane or planes, yet we may regard it as concentrated at a certain point called the centre of pressure. Further, we all know from experience that a pendulum or other suspended body is only still when its centre of gravity is exactly under the point of suspension. If we move it to either side it will swing back again.

In just the same way, the only position in which an aeroplane will remain steady is that in which the centre of gravity is exactly under the point of suspension or, in other words, the centre of pressure. For the centre of pressure in the aeroplane is precisely similar to the point of suspension of a pendulum.

Let us, then, picture to ourselves an aeroplane flying along on a horizontal course with this happy state of things prevailing. Something we will suppose occurs to upset it with the result that it begins to dive downwards. It is then in the position of sliding downhill and instantly its speed increases in consequence. That increase of speed causes the air to press a little more strongly than it did before upon the front edge of the planes. In other words, the centre of pressure shifts forward a little, with the result that the centre of gravity is then a little to the rear of the centre of pressure.

A moment's reflection will show that with the centre of pressure (or point of suspension) in advance of the centre of gravity there is a tendency for the machine to turn upwards again, or, in other words, to right itself.

If, on the other hand, the initial upset causes it to shoot upwards the speed instantly falls off and the centre of pressure retreats, turning the machine downwards once more. And the same principle applies whatever the disturbance may be. Instantly and automatically a turning force comes into play which tends to check and ultimately to correct what has gone wrong.

This principle explains the behaviour of the card dropped from an upstairs window and, no doubt, as has been said, it operated also in the early flying machines, but in their case other factors caused disturbing elements with which the self-righting tendency was not strong enough to cope. As time went on, however, experience taught the makers how to avoid these disturbing factors until at last the self-righting tendency was able to act effectively, thus producing the aeroplane which is inherently stable and which will, for short periods at all events, fly safely without attention from its pilot.

Each little improvement in this direction was an invention. Of course, there were certain men whose names stand out prominently in the history of the aeroplane, notable among whom are the Wright brothers, but the final result is due to innumerable inventions, many of them by unknown men.

But perhaps someone will say, how can you possibly talk about final results in a matter which is still in its infancy?

The answer to that is that so far as the safe, flyable machine is concerned, it has arrived. Little now remains to be done in that direction. Further improvements there will, of course, be, but the great fundamental problems of flight have been solved.

CHAPTER 25

The Aerial Lifeboat

Balloons had not long been invented when the idea arose of a device by means of which an aeronaut who found himself in difficulties might be able to reach the ground in safety. In other words, the need was felt for something which should play towards the balloon the part which the lifeboat does to the ship.

The original idea of a parachute was even older than that, since we are told of a man away back in the seventeenth century who amused the King of Siam by jumping from a height and steadying his descent by means of a couple of umbrellas. It was not, however, until the very end of the eighteenth century or the beginning of the nineteenth that descents were made from really considerable heights from balloons.

The usual arrangement then was to have the parachute hanging at full length fastened below the basket, or tied to one side of the balloon in such a manner that it could be detached by cutting the cords that held it up. When the parachute was carried below the balloon basket the man was already in the cradle or seat of the parachute ready to be dropped, but when the seat was tied to the side of the car of the balloon the aeronaut, when he wished to make a descent, first got from the car into the seat, and, casting himself adrift from the car, swung out from under the centre of the balloon so that when he was hanging clear another man in the balloon cut the cords or pulled a slip-knot which set the parachute free. There were different ways of doing this and when a man was by himself he had to get into the sling of the parachute and, on finding himself clear of everything, he would give a tug to a cord which would release a catch holding up the parachute and allow it to drop to earth.

The parachute, at the very first, was but a simple affair, being little more than a circular sheet of cotton or similar fabric, but it was very soon found necessary to make it *a bag* or it would not properly hold

187

the air. Cords were attached at regular intervals all around the edge of this bag, these cords being gathered together and attached to the edge of a basket which carried the man. Sometimes only a sling was used, or a simple light seat after the fashion of the bosun's chair upon which a sailor is sometimes hauled to the top of an unclimbable mast, or a steeplejack to the top of a chimney.

Thus, when it was dropped, the weight of the man, pulling upon all the cords simultaneously, drew down the edge of the bag, which, catching the air in its fall, acted as a powerful brake and reduced the rate of falling to such an extent that if all went well the man alighted in safety if not comfort.

As has already been remarked in another chapter, air, which seems to us sometimes to be so exceedingly light as to have practically no weight at all, really has weight and also the property which we call inertia, by virtue of which things at rest prefer to stay at rest.

Now when this open air-bag, of considerable area, is pulled down-wards it causes a very considerable disturbance in the air. As it de-scends the air inside and beneath it is first pushed downwards and compressed a little, then it commences to move outwards, towards the edge, round which it finally escapes to fill the slight vacuum in the space just above the descending parachute. All this the air objects to do because of its inertia. The parachute has to force it to act thus and in that way it uses up some of the force of gravity which all the time is pulling the man earthwards. In other words, that force, instead of dragging the man downwards at such a speed as to dash him to pieces, is so far employed in churning up the air that what is left only brings him down quite slowly and ends with just a gentle bump. That is the scientific explanation of what happens, although expressed in some-what homely language.

To anyone who thinks of this matter it will be clear that a relatively heavy weight like a man, suspended from a parachute, is like a very delicately poised pendulum, and consequently it is not surprising to hear that the early parachutes oscillated very considerably from side to side, so much so, indeed, that this oscillation became a decided danger, for before the proper shape of the air-bag was found out they sometimes skidded and even turned inside out. It was found, however, at quite an early stage that this instability could be to some extent cured by making a hole right in the centre or crown of the parachute through which the air compressed inside could blow upwards in a powerful jet. At first sight it seems as if this would much weaken the parachute and cause it to descend too quickly, but quite a large hole can be safely made, and to make such a hole is only the same thing as slightly reducing the area and that can be easily remedied by slightly increasing the diameter.

Reading of this many years ago, I have often been puzzled as to why the presence of the hole should have this steadying effect, the explanation given in the old scientific textbook from which I learnt it being obviously very unsatisfactory. Of recent years, however, this subject of parachutes has been very deeply studied by an eminent engineer of London, Mr. E. R. Calthrop, the inventor of the Guardian Angel parachute to which these remarks are leading up, and he has hit upon what is undoubtedly the explanation. He says that the big jet of air shooting upwards through the crown of the parachute forms in effect a rudder which steers the parachute in a straight downward course, just as the rudder guides a boat upon the surface of the water.

It is quite possible that thus far the impression conveyed to the reader's mind is that the parachute and its use are very simple, straightforward matters. One may be inclined to think that it is only necessary to get a circular sheet of fabric, to fasten the cords to it, to connect them to a suitable seat and then to descend from any height at any time in perfect safety. If you make a model from a flat sheet of cotton, then one made like a bag, and drop them with little weights attached from the top window of your house you will see what funny things the air can do. After having tried these little ones, you will begin to suspect that the big parachute is full of waywardness: and, as a matter of fact, until recent years, it has been very largely a delusion and a snare. By its refusal to act and open at the right moment it has sacrificed many lives. Although apparently so simple, there were conditions existing and forces at work which for a century or more had never been properly considered and investigated, and it is only now that we have arrived at a parachute whose certainty of action and general trustworthiness entitle it to be called the lifeboat of the air.

The troubles with the older parachutes were two. First, although often it opened quite quickly, and carried its load as perfectly as could be desired, it sometimes had the habit of delaying its opening, and unless the fall were from a very great height it was unsafe to take the risk, indeed, it sometimes refused to open at all, and the poor parachutist suffered a fearful death. It had to be carried in a more or less folded-up state. Often it was hung up by its centre to the side of a balloon, when it was very like a shut-up umbrella. Consequently the power of opening quickly and certainly was of the first importance, and the lack of that power and the uncertainty of its action were a very serious defect. It has always suffered from an ill reputation as to reliability.

The second fault lay with the cords. They would persist in getting entangled. Everyone knows how a dozen cords hanging near together will get entangled with each other on the slightest provocation. Such cords if blown about by a strong wind would be much worse even than when still, and if, as must often be the case with parachutes, they be coiled up, we all know from our own experience that some of them would be almost sure to get knotted and tangled together when, in a sudden emergency, the attempt was made to pull them all out of their coils in a second or two. Just picture to yourself what it means: a dozen coiled cords all close together, themselves all coiled up in loops, suddenly pulled. Something awkward appears almost inevitable. And the result of even one rope going awry may be fatal, for it may prevent the parachute opening out fully, probably giving it a lop-sided form incapable of gripping the air effectually and consequently allowing the unfortunate man to fall with a velocity which means certain death. This second cause of failure to open, through entanglement of cordage, has happened in a number of cases, with fatal results.

So much for the faults of the old primitive parachute. Now let us consider for a moment the urgent need for a parachute which is free from such faults. The man who goes up in a balloon on a Saturday afternoon feels so sure of his craft that he thinks he needs no life-boat, yet men in ordinary free balloons have been killed for want of them. The spectators at country fairs no longer appreciate a parachute descent as a great and extraordinary spectacle. But in warfare, with kite balloons by the dozen, with dirigible balloons by the score and aeroplanes by the hundred, the call for parachutes is urgent and irre-sistible. At all events, Mr. Calthrop found an irresistible call to devote years of close study, unceasing toil and considerable sums of money to the task of perfecting an improved parachute which would always open and open quickly, and whose cords would never get entangled. He has the satisfaction of knowing that by so doing he has provided an appliance that in the air is as reliable as a lifeboat is at sea, and that at all times, and from every kind of aircraft, can be depended upon in case of accident to save the lives of gallant airmen who but for his work would be dashed to death. The Great War has taught us to re-gard life somewhat cheaply. For years we were more concerned with taking life than with saving it, yet surely to save the life of one's own men is equivalent to taking the lives of one's opponents, so that even from the point of view of warfare the saving of life may be a help towards victory. This is particularly so when the lives saved are those of the choicest spirits, and among the most highly trained. It has been reckoned that to make a fully-trained pilot costs as much as £1500, so that to save but a few, even in their preparatory nights on the training-grounds where so many accidents happen, makes quite an appreciable difference in the cost of a war, without considering the main question of the men's lives.

Many inventions arise through a man thinking of an idea and then seeking and finding some application for it. Elsewhere in this book, I give examples of such cases. Here we have an instance of the opposite, for Mr. Calthrop found his thoughts strongly directed in this direction by the death of a personal friend, the Hon. C. S. Rolls, one of the early martyrs in the cause of aviation, not to mention others who shared the same risks and in some cases the same fate. His interest thus aroused, he first studied all the records which could be found relating to parachute accidents, so as to ascertain, if possible, what were the causes of failure. Then he commenced a long series of experiments with a view to removing these causes. Improvement after improvement was tried, unexpected difficulties were discovered and grappled with, the kinematograph was called in to record the movements of the falling objects, a task for which it is far better fitted than the human eye, and after years of this there emerged the finished parachute, automatic in its action, perfectly reliable and a true safeguard, which I am about to describe.

The parachute's body consists of the finest quality silk carefully cut into gussets of such a shape that when sewn together somewhat after the manner of the cover of an umbrella, they form a shallow bag, parabolic in section, of that particular shape which the material would assume naturally were it perfectly elastic when enclosing its resisting body of compressed air.

At intervals round the edge are fastened twenty-four V-shaped tapes. These are only a few feet long and the lower end of each V-shaped pair is attached to a long main tape. There are twelve of these main tapes, and their lower ends unite in a metal disc from which is suspended the sling and harness by which the man is supported.

So the twenty-four short tapes form twelve V's to the points of which are attached the twelve long tapes which support the man. The reason why tapes are used in this particular parachute and not cords will be referred to later.

In the crown of the silk body there is the usual hole for the purpose of forming the air-rudder to steady the parachute in its descent.

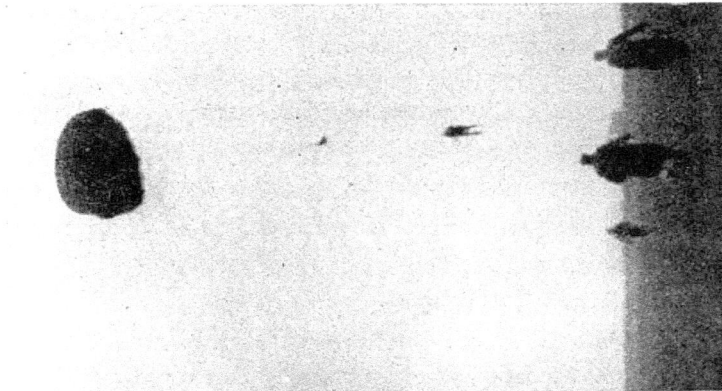

THE GUARDIAN ANGEL PARACHUTE

(1) SHOWS THE AIRMAN IN THE HARNESS BY WHICH HE IS ATTACHED TO THE PARACHUTE. BY MEANS OF THE STAR-SHAPED BUCKLE HE CAN INSTANTLY RELEASE HIMSELF. (2) SHOWS THE PARACHUTE TWO SECONDS AFTER THE AIRMAN HAS JUMPED FROM THE AEROPLANE. IN (3) HE IS SEEN NEARING THE GROUND.

And now we can consider the first great feature of this wonderful invention and ask ourselves these questions: By what means is it made to open? What makes it more reliable than others?

To answer that we must first see why the others sometimes refused to open. In whatever way an ordinary parachute may be packed it must, when coming into use, assume the state of a shut umbrella with a hole in the top.

In this condition it is assumed that as it falls the air will find a way in through the lower end and will blow the parachute open in precisely the same way that a strong wind will sometimes blow out the folds of an umbrella.

But, as a matter of fact, the loose folds of a parachute, when the edge of the gussets is gathered in, are sure to overlap and enfold each other more or less. Thus, when in the shut-umbrella state, it sometimes happens that air which is inside can escape upwards through the hole more easily than fresh air can get in from below. The parachute, in such a state, is, let us imagine, falling rapidly through the air. The result is just the same as if it were still and the air were rushing upwards past it. And the upward rush past the top hole tends to *suck air out* through the hole faster than fresh air can find a way in at the bottom.

This is the principle of the ejector, which engineers have put to many uses. For example, the vacuum brakes employed on many large railways owe all their power to stop a train to a vacuum caused by an ejector. There is a short tube or nozzle, placed in the centre of another tube through which steam blows. The action of the steam in the outer tube as it rushes past the end of the inner tube drags after it the air which is in the inner tube so effectively as to produce quite a good vacuum. And in precisely the same way, the upward rush of air past the parachute, or what is just the same, the falling of the parachute through stationary air, can suck the air from inside the latter and create a vacuum in it if the gussets gathered together at the mouth unfortunately overlap one another and are thus locked together by the pressure of the air striving to get in. Thus, instead of the downward fall causing the ordinary parachute to open, as in most cases it will do quite well, the fall under these particular conditions actually binds its folds together and prevents it from opening. It is true this does not often happen, but the risk is *always* present at every drop, and this unreliability has cost the lives of brave men and women, and the knowledge of this constant risk has led others to write down the parachute a failure, by reason of its known unreliability to open instantly. Even when it does open the depth it falls before it opens is so variable, by reason of the fight between vacuum and pressure, that it may be one hundred feet one time and one thousand feet next time with the same parachute.

Now the Guardian Angel is designed so that those conditions cannot occur. Its silken covering is first laid out on the ground and into the centre is introduced a beautifully-designed disc of aluminium, somewhat like a large inverted saucer, of exceeding lightness but of ample strength for what it has to do. Then the silk body is pleated and folded back over the upper part of this launching-disc and gradually packed so that it occupies but a very small space upon the upper surface of the disc. It is so folded that its edge comes in the topmost layer and also in such a manner that on the tapes being pulled the silk unfolds easily and regularly, flowing down as it were over the edge of the disc almost as water flows if allowed to fall from a tap upon the centre of an inverted saucer. After the folding is complete another aluminium disc is placed above the packed silk body which shields it from the enormous air pressure when it is being released from an aeroplane flying at top speed. The upper and lower fabric covers are then superimposed and sealed and the Guardian Angel parachute is ready for use.

The tapes, likewise, are folded up, in a special way upon the bottom cover, which is sprung over the bottom of the disc. The bottom cover with the tapes upon it, is pulled away by the weight of the airman as he makes his jump to safety, and the tapes are so arranged that a pull upon them causes them to draw out steadily and smoothly, almost like water falling from a height.

If we regard the silk as forming a shallow bag inverted, we may say that it is folded upon the disc inside out and the function of the disc is to cause it to spread and enclose a wide column of air as it is pulled from its folds. To commence with it is nothing more than so much folded-up silk, but from the first moment of action it becomes a bag with a wide-open mouth, for its open mouth cannot be smaller than the disc. Therefore, from the first instant it begins to grip the air and the ejector action never gets a chance to commence. The pressure of air inside is from the very commencement of the fall greater than that of the surrounding air. Moreover, the disc covers the hole until the parachute is actually open, thereby making ejector action doubly impossible.

The widely-opened mouth of the air-bag (I cannot help repeating that term for it is so expressive) swallows up more and more air as the thing falls rapidly, with the result that the air inside is instantly compressed and the increasing pressure as the silk is more and more fully drawn out causes it to expand until the whole is fully extended like a huge umbrella. The instant compression of the enclosed column of air is what causes it *always* to open automatically.

When once it is pointed out it is easy to see what a difference the presence of this disc makes. It is so simple that it cannot fail to act and having once produced that open mouth all the rest is due to the action of natural forces which can be absolutely relied upon. The ordinary parachute with its hopeless irregularities has, in fact, been converted into a machine whose action can *never* fail.

The disc is fastened to the balloon or aeroplane and is left behind when the parachute falls, having done its work.

And now let us consider the tapes. As has already been remarked, a series of coiled cords cannot be relied upon to pull out straight without possibility of entanglement, but a tape, if folded to and fro like a Chinese cracker, will invariably do so. So packed tapes have been substituted for coiled corded rigging, with the certainty that they cannot be entangled in the fiercest air current.

And now we come to another interesting feature. The man is not suspended directly from the small disc to which the tapes are attached but by a non-spinning sling which contains a shock absorber. This latter consists of a number of strands of rubber and it is owing to its action that the aviator who trusts his life to the parachute suffers little or no shock; even when the instant opening of the parachute begins to arrest his fall. And not only does it save him from shock, but it also avoids the possibility of too great a stress coming suddenly upon the parachute or its rigging of tapes.

The aviator himself is attached to the parachute through the shock-absorber sling, by means of a harness which he wears constantly throughout his flight, so that in the event of trouble he only has to jump overboard and the parachute automatically does the rest. This harness consists of two light but strong aluminium tubular rings through which he places his arms, combined with a series of straps which can be so adjusted that the stress of carrying him comes upon those parts of his body best adapted to bear it.

This improved parachute is the only one which is capable of being used instantly and without preparation for descent from an aeroplane flying at top speed. It is easy to see that it is one thing to drop from a stationary or nearly stationary balloon and quite another to dive from an aeroplane at one hundred miles per hour. The latter is equivalent to suddenly trusting oneself to a parachute *during the strongest gale*. It has been found, by experiment, however, that high speed is no bar to the use of this parachute since it only causes the parachute to open a little more quickly than usual, which means that it can be used with safety from an even lower height.

Under the worst conditions this wonderful parachute can be relied upon always to open and carry its load at a height of only one hundred feet, and its use is safe in all circumstances when dropped from two hundred feet above the ground. After it has once got into operation and taken charge of affairs, so to speak, the man descends at the rate of only fifteen feet per second, which is just about the same as dropping from a height of a little over three feet. In other words, he will arrive on the ground with no worse bump than you would get by jumping off the dining-room table.

But suppose that there were a wind blowing: would not the parachute come down in a slanting direction and then drag the man along? Or may he not alight upon a tree or the roof of a house, only to be pulled off again and flung headlong? Quite true he might, were not proper provision made for such occurrences. Embodied in the harness is a lock which can be instantly undone, by a simple movement of a lever in the hand, and by its aid the man on touching earth or on alighting upon anything solid can release himself instantly, after which the parachute can sail away whither it will, but he will be safe and sound.

What Mr. Calthrop has accomplished by the invention of his Guardian Angel parachute may be summarised briefly by saying that he has reduced the minimum height from which a parachute could be dropped from two thousand to two hundred feet, and that he has made it possible to launch a parachute, with the certainty of safety, from any kind of aircraft flying at the slowest or highest speed of which they are capable.

LEONAUR

ALSO FROM LEONAUR
AVAILABLE IN SOFTCOVER OR HARDCOVER WITH DUST JACKET

THE FALL OF THE MOGHUL EMPIRE OF HINDUSTAN by H. G. Keene—By the beginning of the nineteenth century, as British and Indian armies under Lake and Wellesley dominated the scene, a little over half a century of conflict brought the Moghul Empire to its knees.

LADY SALE'S AFGHANISTAN by Florentia Sale—An Indomitable Victorian Lady's Account of the Retreat from Kabul During the First Afghan War.

THE CAMPAIGN OF MAGENTA AND SOLFERINO 1859 by Harold Carmichael Wylly—The Decisive Conflict for the Unification of Italy.

FRENCH'S CAVALRY CAMPAIGN by J. G. Maydon—A Special Correspondent's View of British Army Mounted Troops During the Boer War.

CAVALRY AT WATERLOO by Sir Evelyn Wood—British Mounted Troops During the Campaign of 1815.

THE SUBALTERN by George Robert Gleig—The Experiences of an Officer of the 85th Light Infantry During the Peninsular War.

NAPOLEON AT BAY, 1814 by F. Loraine Petre—The Campaigns to the Fall of the First Empire.

NAPOLEON AND THE CAMPAIGN OF 1806 by Colonel Vachée—The Napoleonic Method of Organisation and Command to the Battles of Jena & Auerstädt.

THE COMPLETE ADVENTURES IN THE CONNAUGHT RANGERS by William Grattan—The 88th Regiment during the Napoleonic Wars by a Serving Officer.

BUGLER AND OFFICER OF THE RIFLES by William Green & Harry Smith—With the 95th (Rifles) during the Peninsular & Waterloo Campaigns of the Napoleonic Wars.

NAPOLEONIC WAR STORIES by Sir Arthur Quiller-Couch—Tales of soldiers, spies, battles & sieges from the Peninsular & Waterloo campaings.

CAPTAIN OF THE 95TH (RIFLES) by Jonathan Leach—An officer of Wellington's sharpshooters during the Peninsular, South of France and Waterloo campaigns of the Napoleonic wars.

RIFLEMAN COSTELLO by Edward Costello—The adventures of a soldier of the 95th (Rifles) in the Peninsular & Waterloo Campaigns of the Napoleonic wars.

LEONAUR

ALSO FROM LEONAUR
AVAILABLE IN SOFTCOVER OR HARDCOVER WITH DUST JACKET

BUGEAUD: A PACK WITH A BATON *by Thomas Robert Bugeaud*—The Early Campaigns of a Soldier of Napoleon's Army Who Would Become a Marshal of France.

WATERLOO RECOLLECTIONS *by Frederick Llewellyn*—Rare First Hand Accounts, Letters, Reports and Retellings from the Campaign of 1815.

SERGEANT NICOL *by Daniel Nicol*—The Experiences of a Gordon Highlander During the Napoleonic Wars in Egypt, the Peninsula and France.

THE JENA CAMPAIGN: 1806 *by F. N. Maude*—The Twin Battles of Jena & Auerstadt Between Napoleon's French and the Prussian Army.

PRIVATE O'NEIL *by Charles O'Neil*—The recollections of an Irish Rogue of H. M. 28th Regt.—The Slashers—during the Peninsula & Waterloo campaigns of the Napoleonic war.

ROYAL HIGHLANDER *by James Anton*—A soldier of H.M 42nd (Royal) Highlanders during the Peninsular, South of France & Waterloo Campaigns of the Napoleonic Wars.

CAPTAIN BLAZE *by Elzéar Blaze*—Life in Napoleons Army.

LEJEUNE VOLUME 1 *by Louis-François Lejeune*—The Napoleonic Wars through the Experiences of an Officer on Berthier's Staff.

LEJEUNE VOLUME 2 *by Louis-François Lejeune*—The Napoleonic Wars through the Experiences of an Officer on Berthier's Staff.

CAPTAIN COIGNET *by Jean-Roch Coignet*—A Soldier of Napoleon's Imperial Guard from the Italian Campaign to Russia and Waterloo.

FUSILIER COOPER *by John S. Cooper*—Experiences in the 7th (Royal) Fusiliers During the Peninsular Campaign of the Napoleonic Wars and the American Campaign to New Orleans.

FIGHTING NAPOLEON'S EMPIRE *by Joseph Anderson*—The Campaigns of a British Infantryman in Italy, Egypt, the Peninsular & the West Indies During the Napoleonic Wars.

CHASSEUR BARRES *by Jean-Baptiste Barres*—The experiences of a French Infantryman of the Imperial Guard at Austerlitz, Jena, Eylau, Friedland, in the Peninsular, Lutzen, Bautzen, Zinnwald and Hanau during the Napoleonic Wars.

LEONAUR

ALSO FROM LEONAUR
AVAILABLE IN SOFTCOVER OR HARDCOVER WITH DUST JACKET

LIFE IN THE ARMY OF NORTHERN VIRGINIA *by Carlton McCarthy*—The Observations of a Confederate Artilleryman of Cutshaw's Battalion During the American Civil War 1861-1865.

HISTORY OF THE CAVALRY OF THE ARMY OF THE POTOMAC *by Charles D. Rhodes*—Including Pope's Army of Virginia and the Cavalry Operations in West Virginia During the American Civil War.

CAMP-FIRE AND COTTON-FIELD *by Thomas W. Knox*—A New York Herald Correspondent's View of the American Civil War.

SERGEANT STILLWELL *by Leander Stillwell* —The Experiences of a Union Army Soldier of the 61st Illinois Infantry During the American Civil War.

STONEWALL'S CANNONEER *by Edward A. Moore*—Experiences with the Rockbridge Artillery, Confederate Army of Northern Virginia, During the American Civil War.

THE SIXTH CORPS *by George Stevens*—The Army of the Potomac, Union Army, During the American Civil War.

THE RAILROAD RAIDERS *by William Pittenger*—An Ohio Volunteers Recollections of the Andrews Raid to Disrupt the Confederate Railroad in Georgia During the American Civil War.

CITIZEN SOLDIER *by John Beatty*—An Account of the American Civil War by a Union Infantry Officer of Ohio Volunteers Who Became a Brigadier General.

COX: PERSONAL RECOLLECTIONS OF THE CIVIL WAR--VOLUME 1 *by Jacob Dolson Cox*—West Virginia, Kanawha Valley, Gauley Bridge, Cotton Mountain, South Mountain, Antietam, the Morgan Raid & the East Tennessee Campaign.

COX: PERSONAL RECOLLECTIONS OF THE CIVIL WAR--VOLUME 2 *by Jacob Dolson Cox*—Siege of Knoxville, East Tennessee, Atlanta Campaign, the Nashville Campaign & the North Carolina Campaign.

KERSHAW'S BRIGADE VOLUME 1 *by D. Augustus Dickert*—Manassas, Seven Pines, Sharpsburg (Antietam), Fredricksburg, Chancellorsville, Gettysburg, Chickamauga, Chattanooga, Fort Sanders & Bean Station.

KERSHAW'S BRIGADE VOLUME 2 *by D. Augustus Dickert*—At the wilderness, Cold Harbour, Petersburg, The Shenandoah Valley and Cedar Creek..

LEONAUR

ALSO FROM LEONAUR
AVAILABLE IN SOFTCOVER OR HARDCOVER WITH DUST JACKET

ESCAPE FROM THE FRENCH *by Edward Boys*—A Young Royal Navy Midshipman's Adventures During the Napoleonic War.

THE VOYAGE OF H.M.S. PANDORA *by Edward Edwards R. N. & George Hamilton, edited by Basil Thomson*—In Pursuit of the Mutineers of the Bounty in the South Seas—1790-1791.

MEDUSA *by J. B. Henry Savigny and Alexander Correard and Charlotte-Adélaïde Dard* —Narrative of a Voyage to Senegal in 1816 & The Sufferings of the Picard Family After the Shipwreck of the Medusa.

THE SEA WAR OF 1812 VOLUME 1 *by A. T. Mahan*—A History of the Maritime Conflict.

THE SEA WAR OF 1812 VOLUME 2 *by A. T. Mahan*—A History of the Maritime Conflict.

WETHERELL OF H. M. S. HUSSAR *by John Wetherell*—The Recollections of an Ordinary Seaman of the Royal Navy During the Napoleonic Wars.

THE NAVAL BRIGADE IN NATAL *by C. R. N. Burne*—With the Guns of H. M. S. Terrible & H. M. S. Tartar during the Boer War 1899-1900.

THE VOYAGE OF H. M. S. BOUNTY *by William Bligh*—The True Story of an 18th Century Voyage of Exploration and Mutiny.

SHIPWRECK! *by William Gilly*—The Royal Navy's Disasters at Sea 1793-1849.

KING'S CUTTERS AND SMUGGLERS: 1700-1855 *by E. Keble Chatterton*—A unique period of maritime history-from the beginning of the eighteenth to the middle of the nineteenth century when British seamen risked all to smuggle valuable goods from wool to tea and spirits from and to the Continent.

CONFEDERATE BLOCKADE RUNNER *by John Wilkinson*—The Personal Recollections of an Officer of the Confederate Navy.

NAVAL BATTLES OF THE NAPOLEONIC WARS *by W. H. Fitchett*—Cape St. Vincent, the Nile, Cadiz, Copenhagen, Trafalgar & Others.

PRISONERS OF THE RED DESERT *by R. S. Gwatkin-Williams*—The Adventures of the Crew of the Tara During the First World War.

U-BOAT WAR 1914-1918 *by James B. Connolly/Karl von Schenk*—Two Contrasting Accounts from Both Sides of the Conflict at Sea D uring the Great War.

LEONAUR

ALSO FROM LEONAUR
AVAILABLE IN SOFTCOVER OR HARDCOVER WITH DUST JACKET

IRON TIMES WITH THE GUARDS *by An O. E. (G. P. A. Fildes)*—The Experiences of an Officer of the Coldstream Guards on the Western Front During the First World War.

THE GREAT WAR IN THE MIDDLE EAST: 1 *by W. T. Massey*—The Desert Campaigns & How Jerusalem Was Won---two classic accounts in one volume.

THE GREAT WAR IN THE MIDDLE EAST: 2 *by W. T. Massey*—Allenby's Final Triumph.

SMITH-DORRIEN *by Horace Smith-Dorrien*—Isandlwhana to the Great War.

1914 *by Sir John French*—The Early Campaigns of the Great War by the British Commander.

GRENADIER *by E. R. M. Fryer*—The Recollections of an Officer of the Grenadier Guards throughout the Great War on the Western Front.

BATTLE, CAPTURE & ESCAPE *by George Pearson*—The Experiences of a Canadian Light Infantryman During the Great War.

DIGGERS AT WAR *by R. Hugh Knyvett & G. P. Cuttriss*—"Over There" With the Australians by R. Hugh Knyvett and Over the Top With the Third Australian Division by G. P. Cuttriss. Accounts of Australians During the Great War in the Middle East, at Gallipoli and on the Western Front.

HEAVY FIGHTING BEFORE US *by George Brenton Laurie*—The Letters of an Officer of the Royal Irish Rifles on the Western Front During the Great War.

THE CAMELIERS *by Oliver Hogue*—A Classic Account of the Australians of the Imperial Camel Corps During the First World War in the Middle East.

RED DUST *by Donald Black*—A Classic Account of Australian Light Horsemen in Palestine During the First World War.

THE LEAN, BROWN MEN *by Angus Buchanan*—Experiences in East Africa During the Great War with the 25th Royal Fusiliers—the Legion of Frontiersmen.

THE NIGERIAN REGIMENT IN EAST AFRICA *by W. D. Downes*—On Campaign During the Great War 1916-1918.

THE 'DIE-HARDS' IN SIBERIA *by John Ward*—With the Middlesex Regiment Against the Bolsheviks 1918-19.

LEONAUR

ALSO FROM LEONAUR
AVAILABLE IN SOFTCOVER OR HARDCOVER WITH DUST JACKET

THE 9TH—THE KING'S (LIVERPOOL REGIMENT) IN THE GREAT WAR 1914 - 1918 *by Enos H. G. Roberts*—Mersey to mud—war and Liverpool men.

THE GAMBARDIER *by Mark Severn*—The experiences of a battery of Heavy artillery on the Western Front during the First World War.

FROM MESSINES TO THIRD YPRES *by Thomas Floyd*—A personal account of the First World War on the Western front by a 2/5th Lancashire Fusilier.

THE IRISH GUARDS IN THE GREAT WAR - VOLUME 1 *by Rudyard Kipling*—Edited and Compiled from Their Diaries and Papers—The First Battalion.

THE IRISH GUARDS IN THE GREAT WAR - VOLUME 1 *by Rudyard Kipling*—Edited and Compiled from Their Diaries and Papers—The Second Battalion.

ARMOURED CARS IN EDEN *by K. Roosevelt*—An American President's son serving in Rolls Royce armoured cars with the British in Mesopatamia & with the American Artillery in France during the First World War.

CHASSEUR OF 1914 *by Marcel Dupont*—Experiences of the twilight of the French Light Cavalry by a young officer during the early battles of the great war in Europe.

TROOP HORSE & TRENCH *by R.A. Lloyd*—The experiences of a British Lifeguardsman of the household cavalry fighting on the western front during the First World War 1914-18.

THE EAST AFRICAN MOUNTED RIFLES *by C.J. Wilson*—Experiences of the campaign in the East African bush during the First World War.

THE LONG PATROL *by George Berrie*—A Novel of Light Horsemen from Gallipoli to the Palestine campaign of the First World War.

THE FIGHTING CAMELIERS *by Frank Reid*—The exploits of the Imperial Camel Corps in the desert and Palestine campaigns of the First World War.

STEEL CHARIOTS IN THE DESERT *by S. C. Rolls*—The first world war experiences of a Rolls Royce armoured car driver with the Duke of Westminster in Libya and in Arabia with T.E. Lawrence.

WITH THE IMPERIAL CAMEL CORPS IN THE GREAT WAR *by Geoffrey Inchbald*—The story of a serving officer with the British 2nd battalion against the Senussi and during the Palestine campaign.

www.ingramcontent.com/pod-product-compliance
Lightning Source LLC
Chambersburg PA
CBHW032056080426

42733CB00006B/292